QL687.AIS576 1987

Dr. James R. Karr

Skutch, Alexander F. **A Naturalist Amid Tropical Splendor.**
Univ. of Iowa Pr. Apr. 1987. c.230p. illus. by Dana Gardner. bibliog. index. ISBN 0-87745-163-X. $22.50. NAT HIST

For over 50 years Skutch, the dean of neotropical ornithology, has sprinkled his natural history writings with his philosophy of nature and man. This volume is divided between natural history of tropical organisms (mostly birds) and his perspectives on subjects such as the origin of beauty, ethics, and feeling close to nature. His descriptions of events in nature are accurate but his interpretation of those events and the role of humans in nature stems more from his idealistic philosophy than from reality. For example, he counsels against conservationists saving "savage and destructive" predators. Recommended, with reservations, for collections serving lay readers and specialists.—*James R. Karr, Smithsonian Tropical Research Inst., Balboa, Panama*

This is an uncorrected proof of a review scheduled for Library Journal, Apr. 1 1987

A NATURALIST
AMID TROPICAL SPLENDOR

A NATURALIST AMID TROPICAL SPLENDOR

BY ALEXANDER F. SKUTCH
ILLUSTRATIONS BY DANA GARDNER

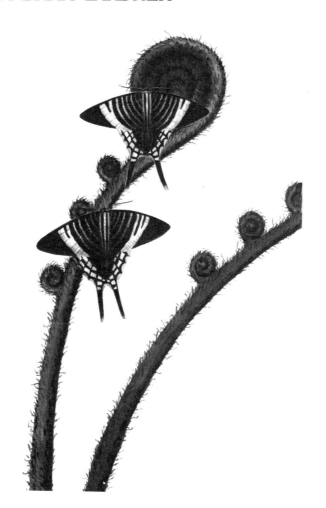

University of
Iowa Press

Iowa City

University of Iowa Press, Iowa City 52242
Copyright © 1987 by the University of Iowa
All rights reserved
Printed in the United States of America
First edition, 1987

Book and jacket design by Patrick Hathcock
Typesetting by G&S Typesetters, Austin, Texas
Printing and binding by Malloy Lithographing, Ann Arbor, Michigan

Library of Congress Cataloging-in-Publication Data

Skutch, Alexander Frank, 1904–
　A naturalist amid tropical splendor.

　Bibliography: p.
　Includes index.
　1. Birds—Latin America—Behavior.　2. Birds—Tropics—Behavior.　3. Natural history—Latin America.　4. Natural history—Tropics.　I. Title.
QL687.A1S576　1987　　598.298　　86-25079
ISBN 0-87745-163-X

Title page: diurnal moths, *Urania fulgens,* on unrolling fern frond

No part of this book may be reproduced or utilized in any form or by any means, electronic or mechanical, including photocopying and recording, without permission in writing from the publisher.

Contents

Preface   vii
1. The Foundations of Tropical Splendor   1
2. A Valiant Dove   6
3. The Birth of Beauty   14
4. Castlebuilders   23
5. Parental Care and Family Life in Birds and Mammals   40
6. A Charming Thief   47
7. On Feeling Close to Nature   52
8. Curious Flowers   58
9. The Appreciation of Nature as a Unifying Power   68
10. A Village in the Treetop   74
11. A Perennial Aspiration   85
12. The Oriole-Blackbird   94
13. A Difficult Choice   101
14. The Scaly-breasted Hummingbird   106
15. Our Animal Heritage   119
16. A Bird of Stormy Heights   127
17. Musings at Birds' Nests   142
18. Versatile Leaves   148
19. The Convivial Ascetic   156
20. A South American Marsh Bird   160
21. The Message of Birds   167
22. A Mockingbird in Blue   172
23. The Lonely Vanguard   181
24. The Elusive Queo   188
25. The Growth of Caring   192
26. The Smallest Tanagers   199
27. The Emotions of Parent Birds   211

Epilogue   222
Bibliography   224
Index   226

Preface

In the humid tropics, life reveals its constructive power as nowhere else. With continuously favorable temperature, abundant sunshine, and no lack of moisture, plants thrive in greatest profusion, and produce the most diverse forms and the most magnificent floral displays. Birds of many kinds wear the most resplendent plumage, build the greatest variety of nests, and practice the most elaborate courtship rites. Butterflies spread the most gorgeous wings as they flit through balmy air. Wasps, ants, and bees construct the greatest diversity of nests, and vast numbers of less conspicuous organisms, many still unnamed, live obscurely amid all this abundance.

The eager young naturalist who, from the less richly endowed temperate zone, plunges into the midst of tropical splendor may almost be overwhelmed by the multitude of strange and fascinating things clamoring for his attention, to be named and classified, to be studied, to be photographed or drawn. He may find it all too much for him and return to his distant homeland, where he feels more secure, where more knowledge is readily accessible through books and journals and the experience of his colleagues, where the unknown does not so press upon him. Or, rising to the challenge of the unknown, he may drive himself almost to the point of exhaustion to make new discoveries in his special field. Falling irrevocably under the spell of tropical nature, he may remain long years, observing, recording, trying to fill gaps in his knowledge.

As the years slip by, our naturalist may find fewer new things to study and, spending less time in observation, give more to the interpretation of what he has experienced. He ponders the significance of all this restless activity, these so diverse forms, this multitude of competing or cooperating organisms that fill the living world. Long rainy afternoons, nights whose stillness is broken only by the sounds of woods and streams, encourage reading, meditation, and writing. He may develop an interpretation of nature different from that of the professional philosopher whose days are passed largely in academic halls and libraries. Accordingly, in this book I have mingled factual accounts with reflections on what I have experienced. I tell about some of the most fascinating of the birds which have been my chief study and some of the unusual plants that surround me. In other chapters, I try to interpret what I have seen during more than half a century amid tropical splendor.

The certainty of our observations contrasts with the fallibility of our interpretations. The earnest student of nature would be highly offended by the colleague who doubted what he claims to have seen clearly and perhaps repeatedly, but his interpretations are open to free discussion. Yet to be content to remain always on the surface, with the phenomena, is to forego the possibility of understanding more deeply or of developing a helpful world view or philosophy of life. Hazardous as our efforts to penetrate the surface of things may be, without them we shall never gain enlightenment. If ever the day arrives when man, fascinated by the discoveries and practical achievements of the sciences, ceases to wonder and speculate about the great, perennial questions that are beyond the purview of our sciences, something precious will have departed from the human spirit.

When the name of an organism is capitalized, the scientific designation is given in the index.

1. *The Foundations of Tropical Splendor*

I look over a wide green landscape flooded with brilliant tropical sunshine, from the clamorous mountain torrent below me to distant craggy summits. Amid the lush verdure glittering hummingbirds and painted butterflies sip nectar from bright flowers, lovely tanagers and great-billed toucans seek colorful fruits, while overhead a flock of green, red, and blue parrots flies noisily against an intensely azure sky. Light brings me awareness of all this splendor, uniting me with it as nothing else can. It reports without delay any visible movement within the wide circle of my vision, for at a velocity of 186,000 miles, or 300,000 kilometers, per second it traverses almost instantaneously the few miles that separate me from the most distant summit.

No less wonderful than light's unique velocity is its ability to travel in straight lines for immense distances. The valley over which I gaze is filled with billions or trillions of rays as closely packed as molecules in the air, crossing one another at every possible angle. At every point of its journey to my eyes, a beam from a leaf across the river intersects with innumerable others; the number of such encounters in a traverse of a hundred yards is inconceivably great. Nevertheless, each ray preserves its individuality amid the throng, contributing to a clear image of its distant source. This capacity of a beam of light to preserve its integrity in a medium densely threaded with other beams, to me one of the most marvelous facts about light, is hardly mentioned in textbooks of physics, which for simplicity treat light rays as though they sped through space without encounters.

The thought of innumerable beams passing from every direction through a point in space without loss of identity makes us ponder what light can be. Although we are told by physicists that it combines properties of particles and waves, it can hardly consist of solid or mutually impervious particles however small, for they would certainly collide and knock each other out of their courses.

There is small probability that a single particle could pass in an undeviating line through a wide space saturated with light, and we should see nothing clearly, if we saw it at all.

The theory that light is propagated in the form of minute waves, or transverse vibrations, of measurable length and frequency neatly explains, by geometrical constructions, the common optical phenomena. Undulatory motion is conceivable only in a medium—solid, liquid, gaseous, or other—capable of vibration. Since light is propagated through a vacuum, whether artificially produced in a bell jar or in outer space, it is obvious that none of the three familiar media is involved. To supply the deficiency, an earlier generation of physicists postulated the luminiferous ether, imponderable and invisible, filling all the space within the Universe. Since it is impossible to separate the ether from space or to find any space through which light cannot pass, they before long recognized that the ether is nothing other than space itself and discarded it as a superfluous entity. Space is the medium that transmits light.

To recognize that space transmits light, which can hardly consist of solid particles, is to be convinced of the untenability of the doctrine, propounded in ancient times by Leucippus and Democritus and widely held in early modern times, that space is simply void extension, the interval between bodies, a container devoid of intrinsic properties. As Plato, Aristotle, and the Stoic philosophers recognized long ago, space is certainly more than that; at the very least, it must have the qualities once ascribed to the ether. But when we consider what is happening continuously at every point of light-filled space, such as that over the valley filled with bright sunshine that I survey, we are perplexed. At any point you choose, innumerable light waves of different frequencies are passing in every direction without losing their identities. Moreover, this same point of intersection is simultaneously traversed by ultraviolet rays, infrared rays, radio waves—almost the whole range of electromagnetic waves. Add to this the magnetic and gravitational fields that permeate it. What kind of medium can at the same instant vibrate in so many different rhythms at a single point, can transmit so many different motions and forces without scrambling them?

To be sure, other media can undulate simultaneously in various frequencies. We can distinguish one voice amid several, hear different sounds at the same time; but the number of sound waves that can pass through the atmosphere without losing identity appears to be limited. The clamor of the mountain torrent, compounded of many different noises made by water rushing over rocks, falls upon my ears as a confused roaring in which, as in the rumble of a great city, I can distinguish no note clearly. But a vastly greater number of light waves, from every visible point in the wide landscape, reaches me with so little loss of identity that, with unaided vision or my binocular, I can distinguish fine details at a great distance.

Nothing is more familiar to us than the space in which we live. It permits us to move freely over a horizontal surface; it accepts whatever we place in it; it is so transparent that it seems to hide no secrets; it is the one thing under the Sun that appears perfectly simple and uncomplicated, easy to understand. We determine its dimensions, whether in a box, a room, the solar system, or the Universe, in whatever units of measurement we choose and are satisfied that we know all that is to be said about it, although actually we do not begin to understand it. So tenuous that it does not resist the movement of a feather, space is nevertheless so strong that it holds Earth in its orbit around the Sun and the Moon in its course around Earth. Binding all things together by means of its unexplained property that we call gravitation, transmitting electromagnetic waves and other influences over vast distances, it makes a coherent Universe of bodies which, if space were no more than void extension, would be a loose collection of unrelated objects. If we knew the intimate nature of space, we might understand everything that it contains. Contemplation of the enigma of space should dispel all dogmatism, assertive or negative, and make us duly humble while we are filled with gratitude for the privilege of living in a medium that almost instantaneously and with no loss of identity can transmit the light waves which so enrich our lives.

Electromagnetic waves within the range of visible light transmit most of the energy that the Sun sends to Earth and the other planets, yet for this the capacity of light waves to preserve their individual identity amid innumerable others hardly seems necessary. A massive influx of mingled waves that had lost their original directions might convey much energy and serve for photosynthesis. Some of the properties of light waves—their stubborn preservation of individuality, the capacity of each to indicate the exact direction of its source—seem to become significant only when they fall upon eyes that report to receptive minds, to which they reveal in detail the forms of objects near and far, to which they bring beauty and knowledge and a sense of intimacy with the larger surrounding world. Nevertheless, for long ages after the Sun began to lavish its rays upon Earth, eyes equipped with lenses and sensitive retinas were nowhere to be found, save possibly in inhabitants of planets beyond the solar system who viewed our Sun as one among a multitude of stars.

A world thickly laced with light waves yet devoid of beings able to see would be incomplete, lacking something necessary for its perfection. High values that it potentially contained would remain unrealized. Such a world would be like a stage set for a splendid drama that is never performed. Evidently the Universe does not rest like an unfinished house but is ever striving to complete itself by the full realization of its potentialities for joyous, satisfying existence. The difficulties are immense, the hazards great, but its resources are vast and its time unlimited.

The preparation of our planet for the more advanced forms of life was an immensely long process, the details of which are still being revealed by scientists in many fields. Concentrating attention upon the material aspects of this preparation (temperature, properties of soil and atmosphere, the living environment necessary to support large, air-breathing animals), we often neglect another aspect equally important to man. Conceivably, the planet might supply everything necessary for the physiological processes of a large animal while offering little for its spiritual or intellectual development, as a vat of syrup provides all that yeast cells need for their growth and multiplication but nothing, as far as we can tell, to nourish a spiritual being. On our planet it has been otherwise; while developing an environment fit to support man's body, it has simultaneously prepared itself to nourish his spirit most generously, with the beauty of the heavens, the grandeur of mountains, and countless lovely or stately vegetable and animal forms that foster aesthetic appreciation and challenge his intellect to learn about them. The effectiveness of much of this preparation depended upon the ability of light waves to intersect at all angles without losing their identity or deviating from straight paths. More than all else, light prepared the world to become a fit abode for spiritual beings. Coeval with the Universe, the world had to wait an enormously long age for life to arise and avail itself of the precious benefits offered.

The human eye, like the eyes of many other vertebrate animals, is an excellent optical instrument whose evolution by the natural selection of random variations required countless generations. Eyes are far from being the whole of our visual apparatus, which includes nerves and brain. It is useful to distinguish between the optical image and the visual image. The former is a pattern of vibrations of diverse frequencies focused by the lens on the rods and cones of the retina. It is a greatly reduced, inverted, intrinsically colorless replica of the visual field—a landscape, building, or person. If we saw the world as it is represented in our eyes, it would appear to be very small, upside down, and severely plain. By a marvelous transformation that we are far from understanding, this drab optical image becomes the often richly colored visual image that illuminates our consciousness, showing us in the wide landscape instead of the landscape in us and with everything right side up.

This visual inversion, whereby we see ourselves in the world instead of the world in us, has far-reaching spiritual as well as practical consequences. If the visual image were the same as the optical image, it would be a poor guide to action, possibly making us try to thread a needle or plant a garden that appeared to be inside our heads, and doubtless it was to promote effective action that the visual inversion evolved. Moreover, we might be far more egocentric and aloof from the world if we saw it on a greatly reduced scale inside us instead of projected outward in three-dimensional space, reduced in size only by perspective and arrayed in attractive colors.

A human being might be described as an expansive spirit enclosed in an insulated body. Like any warm-blooded animal, we need insulation, as in skin and clothing, to preserve our body's heat and vital fluids and exclude injurious organisms and substances that may touch us. Our senses save us from the insulation that would oppress our spirits while it safeguards our tissues. Of our five senses, that which most helps our spirits to reach outward and expand is sight. As light waves stream into our eyes, bringing visions of fair and friendly things, our spirits appear to retrace the beams back to their sources in loving communion. Nothing so attaches us to the surrounding world or reconciles us to our precarious human situation as the visions of beauty and grandeur that light conveys to us. We love beauty because it makes Earth appear friendly to us.

Meditation upon light and its counterpart, vision, reconciles us to nature, or the Universe, in a more profound and philosophic sense. Ages before life arose, before even stars and planets condensed from the cosmic dust, the ability of space to transmit light waves in straight lines, undistorted by all the other waves that cross their paths, offered a means of enhancing existence by serving as a vehicle of communication between bodies and creating visual beauty. To realize the high values that light was capable of generating required optical apparatus of good quality reporting to receptive minds. In the absence of a Supreme Intelligence able to create advanced forms of life directly from the elements, eyes and minds could arise only by the blundering methods of organic evolution, a slow, harsh process involving the replacement of more primitive creatures by more advanced creatures. Nevertheless, persisting through inconceivably long ages, nature produced animals whose lives are immeasurably enriched by the values that refined vision—especially color vision—can bring to grateful, appreciative minds.

The world process moves toward the enhancement of existence by the realization of potentialities present in the primal Universe. Nature does not rest so long as attainable values remain unrealized. Ceaselessly recombining the elements in patterns of ever-increasing amplitude, complexity, and coherence, nature advances by a devious course until it produces the structures necessary for the actualization of these values. Without foreseeing guidance, nature's progress is inevitably slow and often painful; but with growing intelligence, insight, and self-control, we should be able to accelerate the progress while we mitigate it. Knowledge that our unquenchable thirst for a happier, more richly rewarding existence is a more pointed expression of a movement that stirred in the primal depths of the Universe, of which we are parts, should encourage us to persevere in our efforts to overcome all the doubts and difficulties that assail us in an overcrowded, overexploited world while it increases our devoted attachment to a planet made beautiful by light.

The splendors that light brings to us are nowhere more profusely and continuously displayed than in the more humid regions of the tropics, where plants grow and bloom and birds wear their brightest plumage throughout the year. Here, too, are the richest still unrealized opportunities to explore nature's secrets. While we delight in the beauty around us or eagerly add to our wealth of knowledge, we may forget that we owe our enjoyment of beauty, no less than our ability to observe and study natural phenomena, to the mysterious medium in which we live and move, to space which can transmit a thousand rays of light through a single point without distorting the least of them.

2. *A Valiant Dove*

This morning, when I threw out corn for the chickens, ten doves, big enough to be called pigeons, gathered on the dewy lawn to share the yellow grains with the domestic fowls. Most were White-tipped Doves, also called White-fronted Doves and here in Costa Rica known as Coliblancas. Their upper plumage, including wings and tail, was grayish brown, paling to pinkish brown on the crown and almost white on the forehead. On their outer tail feathers were broad white tips, conspicuous when the tail was spread in flight, bordered toward the base with black. Their underparts were much lighter, whitish on the throat, grayish on the breast and sides, white on the abdomen. Each bright yellow eye was set amid bare blue skin. Their bills were black and their legs and toes dull red. Only when a dove raised a wing high above its back to threaten another in a contest for a grain of maize did it reveal the brightest color of its plumage, the cinnamon on the underside of its uplifted wing.

Mingled with the White-tipped Doves were several Rufous-naped Doves, similar in size and appearance but with darker upper plumage, less white on the ends of the outer tail feathers, and red instead of blue skin around their yellow eyes. They also had beautiful cinnamon wing linings. The two kinds of doves treated individuals of their own and the related species much the same, warning them with a wing raised like a sail if they came too near but never fighting.

These two doves illustrate extremes of geographic distribution. The Rufous-naped is found only in a limited area on the rainy Pacific side of southern Costa Rica and across the border in the Panamanian province of Chiriquí. The White-tipped ranges over a vast territory from southern Texas to Central Argentina, including every country on the American continents except Canada and Chile. From warm lowlands it extends upward to about 8,500 feet (2,600 meters) in Guatemala and 10,000 feet (3,050 me-

White-tipped Dove incubating in the scrambling fern, Dicranopteris pectinata

ters) in Venezuela. Absent from the West Indies, it lives on islands along the Venezuelan coast, including Trinidad and Tobago.

Over much of their range, White-tipped Doves are most abundant in the drier regions, but here in southern Costa Rica they are numerous in clearings in heavy rain forests. Avoiding the interior of these forests, they frequent plantations of coffee and bananas, pastures, gardens, thickets threaded by a maze of cowpaths, and open groves. Unlike many other pigeons, they never flock but, singly or with a mate, walk with bobbing heads over open ground, where they appear to find all the seeds, berries, and insects that they eat. Their hunger satisfied, they fly up to rest on the limb of a tree, rarely higher than 30 feet (9 meters). Here, well screened by foliage, with puffed-out chest and closed bill, the male repeats his deep, resonant *coo-ooo*. Occasionally he calls more melodiously in a higher, almost soprano voice, *coo-woo* or *cu cu cu coo-ruuu*.

In southern Costa Rica, White-tipped Doves nest chiefly during the drier part of the year, from December or January to April, most frequently in March. In rainy May and June, when the great majority of our birds have nests, we have found none of the White-tipped Doves. In the wet months from July to October they breed again in smaller numbers. Their neighbors the Rufous-naped Doves and the Blue Ground-Doves likewise have two separate breeding seasons; but in other regions, such as Trinidad, White-tipped Doves are reported to nest throughout the year.

For their nests White-tipped Doves choose sites that are usually from 1 to 15 feet (.30 to 4.5 meters) up, rarely on the ground or, amid concealing vegetation, as high as 60 feet (18 meters). They seek firm supports on matted grass, on stumps, in thick forks, on interlaced twigs or tangles of vines in a thicket, amid the dense foliage of an orange tree, or on stout orchid plants growing on trees. Like other pigeons, these doves practice a division of labor that they rarely vary. The female sits on the growing nest, building it with materials that her mate delivers to her. He walks over the ground, picking up straws, twiglets, and rootlets, often dropping those that he finds too far decayed. When the piece is satisfactory, he flies up to the nest with it. He never gathers a whole sheaf of materials, as building songbirds often do, but on each trip to the nest carries only one piece, which may be branched. Usually he stands upon his mate's back while he lays his contribution beside her, but sometimes he deposits it as he passes beside her. As he approaches, she often vibrates the tips of her wings. After he has flown down to seek another piece, she arranges what he has brought, rotating to shape the nest by tucking in pieces all around. As the nest nears completion, the male brings smaller, finer pieces for the lining. After a spell of work, the female may join her mate on the ground, where they touch bills and nibble the feathers of each other's necks.

One nest was built in three or four days. Compared to the exquisitely finished constructions of many birds, pigeons' nests are crudely built. Those of White-tipped Doves are no exception to this rule, although they are often more substantial than the nests of some other species. Shallowly concave platforms, they measure 5 or 6 inches (12.5 or 15 centimeters) in diameter, not including the projecting ends of sticks. Among the 350 pieces in one nest were weed stems, straws, twigs, dry pieces of vines, fragments of fern fronds, and rootlets. The longest stiff pieces were two crooked straws about 12 inches (30 centimeters) long, but a curved length of vine measured, when straightened, 20 inches (51 centimeters). A much slighter nest contained only 109 items, including vines up to 28 inches (71 centimeters) long and grass stems 20 inches long.

Many species in the pigeon family regularly lay only one egg; many others, including the White-tipped Dove, lay two. It is difficult to learn when the female deposits her first egg. She settles on the eggless nest in the evening and flies off next morning leaving an egg which could have been laid in the evening, during the night, or early in the morning. This egg is covered much of the time by one parent or the other until the second is laid on the following morning, more than twenty-four hours after the first. Like most pigeons' eggs, these are plain white or sometimes pale buff.

Incubation follows the pattern widespread in the pigeon family, with the parents replacing each other on the nest only twice daily. At nests where I could distinguish the sexes by the slightly abnormal plumage of one of them, the individual who sat on the nest arranging it during construction always took the night session on the eggs, and, since in pigeons whose sexes differ in coloration this is regularly the female, I concluded that the White-tipped Dove who spent most time on the nest was of this sex. Between three and half-past four in the afternoon she settled on the eggs, to remain until relieved by her mate between nine o'clock and noon next day, usually between ten and eleven at a nest that I watched. He incubated until his partner returned from four to six hours later.

On the last two days of incubation, one male, evidently anticipating the nestlings, came much earlier, and his mate replaced him before midday. In the afternoons of these days, each partner took another turn, with the result that they changed places four times a day instead of the usual two. The changeover was always accomplished silently, without ceremony. The oncoming partner would alight on a branch of the orange tree and wait there until the other rose and walked out along a limb to take off; then the newcomer walked to the nest and settled on the eggs. Unless suddenly alarmed, pigeons do not fly directly from their nests, as hummingbirds and some others do, or alight directly upon them. Such behavior might shake the eggs from shallow nests.

Fourteen days after the second egg is laid, the second nestling hatches. At one nest, the female neglected to remove an empty shell but the male carried off, separately, both its parts. The blind newly hatched doves bear tufts of the straw-colored, hairlike feathers typical of pigeons.

About the time the nestlings hatch, the parents secrete in their crops a curdlike substance called pigeon's milk. Like mammalian milk, its production is stimulated by the hormone prolactin, formed in the anterior lobe of the pituitary gland. Rich in proteins, fat, and ash, it lacks the sugar present in mammalian milk and, in contrast to the situation in mammals, both the male and female parents yield it. At first the nestlings receive almost pure milk, but as they grow older it is mixed with increasing amounts of solid food that the parents have picked up until it forms a minor part of the young pigeons' diet.

When a blind newly hatched dove rises hungrily in front of its brooding parent, the latter takes the nestling's mouth into a corner of its own and regurgitates milk to it. After a few days, nestlings with opening eyes can see to insert their bills into opposite corners of the parent's mouth. Then all three heads bob up and down together as, with what appears to be strenuous muscular effort, the parent pumps up a mixture of milk and solid items.

Unlike songbirds and many others who make a separate visit to the nest with each meal that they deliver, pigeons can feed their nestlings repeatedly from the supply in their crops without going to find more food. This is especially true during the first days after hatching, when, over an interval of several hours, the continuously brooding parent can feed the nestlings each time they rise up demanding more nourishment. As the nestlings grow older, less time is devoted to feeding them, doubtless because food is regurgitated in more copious streams. During its second day of life, the nestling in one of the few single-egg nests of the White-tipped Dove that I have found was fed nine times by its father and twice by its mother for a total of twenty-one minutes. When this nestling was fifteen days old, its father fed it thrice and its mother twice. The five meals together took less than five minutes.

In all of the foregoing, White-tipped Doves differ little from other members of the pigeon family, but no other dove or pigeon that I have watched attends its young so carefully and devotedly. Some species start to foul their nests while they incubate and permit the nestlings' droppings to accumulate until the nest is heavily soiled; White-tipped Doves keep their nests as clean as songbirds do. As long as the nest is occupied they remove all droppings by swallowing them. Sometimes they do this just before they deliver food, making us ask how they avoid returning the nestlings' wastes to them, if they do avoid this. A female brooding her nestling in the rain drank drops from surrounding foliage. Although somewhat exceptional in keeping their nests clean, White-tipped Doves are not unique in the family. Their forest-dwelling neighbors the Ruddy Quail-Doves likewise remove all waste from their very slight nests.

Many birds leave their nestlings alone, by day and even by night, after they are feathered well enough to maintain their body temperature without a parental coverlet. Not so the White-tipped Doves, who accompany their young almost continuously as long as they remain in the nest. After the first ten days, nestlings already fairly well clad in plumage need little brooding on mild days; they stick their heads out in front of their parent's breast or lie exposed in front of or beside it, to retire beneath the adult when rain falls or night approaches. While brooding or guarding nestlings, the parents follow much the same schedule as while they incubated, except that at a nest which I watched carefully the father now came earlier in the morning, usually between seven and eight o'clock, sometimes sooner, and stayed longer, often nearly seven hours. During these long sessions of attendance at the nest the young are fed. One mother was inconsistent in feeding; on some mornings she fed the nestlings before she left and on others she did not. Similarly, when she returned in the afternoon she might or might not deliver a meal.

When fifteen or sixteen days old, the young doves leave the nest, looking much like their parents, although their plumage is duller and the feathers of breast and back have pale margins, giving them a scalelike appearance. Their eyes are brown instead of yellow. On several occasions, as I approached a nest with well-feathered young, the guardian parent dropped to the ground to limp and flutter away—the ruse used by many birds to lure hungry predators from their eggs or nestlings with the deceptive prospect of an easily caught meal—while the young doves flew strongly in different directions. When an animal threatens feathered young, they may owe their lives to a parent's watchful presence at this critical moment.

When eggs or flightless nestlings are present, the parent does not so readily abandon the nest at the approach of danger, real or apparent. One day a boy led me to a nest he had found in a small coffee plantation. Well above our heads in a shade tree, the dove felt so secure that it sat firmly while we shook the supporting branch. Next morning I returned with a long stick and a mirror to see what the nest contained. When I touched the parent's tail with the stick, it raised both wings straight above its back in an attitude of defiance that revealed their beautiful cinnamon undersides. When I persisted in trying to make the dove depart so that I could see what it covered, it struck the stick resoundingly with its right wing.

Since the steadfast parent refused to leave, I tried to see the contents of its nest when it rose up to defy me. Tying my mirror to the stick, I raised it toward the bird, who dropped to the ground as the intruding object came near. Hopping and limping away, quivering its wings or loosely flapping them, the retreating dove so convincingly acted the part of a crippled bird that I feared it had injured a wing when it struck my stick. I followed at a walking pace until the dove, moving over the freshly cleaned ground as though in the greatest agony and distress, had led me 200 feet (61 meters) to the edge of the plantation, where the dense growth of bushes forced it to interrupt its display. Easily flying over the barrier into the thicket, it alighted on a prostrate log in full view and stood fluttering its wings as though trying vainly to fly. Tangled vines prevented my following, and after a minute of this acting the dove vanished amid the dense vegetation. When I returned to the nest, my uplifted mirror reflected a single half-grown nestling.

Some naturalists have thought that a parent bird, torn between attachment to its nest and fear for its own life, flees in a frenzied, uncoordinated manner, perhaps having a convulsion. This dove's display, and similar performances that I have witnessed in many species of birds, has convinced me that injury-feigning parents are in full control of their faculties. If the dove had been distraught it might have blundered into the dense vegetation and become entangled instead of flying over the obstacle to resume its acting when it found a clear space. Few things that birds do require cooler judgment and weighing of risks than these distraction displays. A blundering performance could be fatal to both the parent and its orphaned young, but I have never known a bird to be caught while acting as though crippled.

A few years ago, a pair of White-tipped Doves nested at the edge of the dense tangle of a branching fern on the bank behind our house. The female permitted me to touch her lightly while she sat, as her more timid or distrustful mate never did. One evening in the twilight, I was alerted by a loud sound coming from the fern, as though the female, brooding her nestlings, were striking something with a wing. I could see little amid the dark shadows, so I ran to the house for a flashlight and a stick. The thuds continued, and the beam of light revealed a large snake at the nest. The wiry fern stipes broke the force of the blow that I directed against the serpent, which vanished into the fern. Despite the shaking of the nest during this encounter, the dove remained sitting. Next morning, only one of her two nestlings was present. Whether one was swallowed by the snake or knocked from the nest in the scuffle, I could not learn. Three evenings later, I again heard wing flapping in the twilight but could detect no snake near the nest. Later in the night, I found the valiant dove sitting peacefully on her nest, apparently covering her remaining nestling. By morning, the nest was empty. What a pity, I thought, that a dove so worthy to be chosen as a model of devoted parenthood is so often shot, stoned, and trapped by thoughtless men and boys, in addition to all the other enemies that assail it!

After the young doves fly from the

nest, the parents continue to attend them. After an early brood of two took wing, the family of four roosted amid the dense foliage of an orange tree. On different nights, these doves arranged themselves in diverse ways. One night three were pressed together in a row, with the fourth two inches distant. On other nights they perched two by two, the members of each couple in contact. Like other pigeons, and like hummingbirds, they slept with their heads forward and exposed, not turned back among the feathers of a shoulder, as many birds sleep.

A few days later, in the same orange tree, the parents started a nest 18 feet (5.5 meters) up, one of the highest White-tipped Doves' nests that I have seen. While they built, one of the juveniles came to beg for food but received none. Then for the next hour, while building was suspended, the young doves sat on the nest, sometimes both together, for longer periods one alone while the other rested nearby. The dove on the nest preened and toyed with the materials. On the evening after an egg was laid, a young dove sat on the nest with its breast over its mother. Apparently, it did not remain after nightfall; but on a rainy evening a few days later, mother and daughter sat side by side, as though both were incubating the single egg that had been laid. They stayed on the nest until it was quite dark, while the other juvenile slept alone high in the orange tree. After this, I no longer saw the young of the first brood at this second-brood nest; but for a while the two kept close company, walking over the lawn and pasture together while they sought food.

A recent experience increased my already great respect for the White-tipped Doves' parental competence. In late July, a pair built a nest on top of a recently pruned living fence post 10 feet (3 meters) high amid young green sprouts that screened it well. While incubating and brooding nestlings, the doves sat firmly while we passed through the narrow gate below them. One morning, a man carrying a long bamboo pole for pulling down Pejibaye fruits accidently brushed against the sprouts around the nest, causing the parent to fly off, immediately followed by the two nestlings it was covering. Only ten or eleven days old and partly feathered, the young fluttered to the ground and tried to escape by running. We caught one and replaced it on the nest, where it was brooded, but the other eluded our search.

That evening, as I walked toward the gate, a parent fluttered over the ground for many yards, waving its wings in an excellent distraction display. Near the point where the display began, I found the missing young dove. Now fairly well feathered, it could fly a short distance and run too well to be caught. Next morning, a parent twice gave a distraction display, and in each case I discovered the young dove nearby. In the afternoon, I found the young bird and a parent, possibly brooding it, beneath the privet hedge near the nest. At my approach, the adult flew off and then the young dove, flying fairly well. Alighting, the parent watched where its offspring went, then joined it and fed it. All this while, the nestling on top of the post was brooded much of the time. The parents were attending both members of their prematurely separated brood.

On the next two days, I failed to find the second young dove. Possibly it had returned to the nest, where I could not see it because it was so constantly brooded by a parent whom I did not wish to disturb, thereby risking another premature departure. In any case, on the following day, the fourth since the accidental disruption of the nesting, both young were lying on the nest in front of a guarding parent. They were brooded, or at least accompanied, through much of that day, and on the following morning both left at the usual age of fifteen or sixteen days.

Widespread among birds is the division

of a brood newly emerged from the nest, one parent taking charge of one or more fledglings while the other parent attends the rest of the family. Normally, this happens after the young leave the nest at the usual age and are no longer brooded. With the doves, the situation was more complicated, for although I could not distinguish the parents I have no doubt that both continued to brood on the nest, as pigeons usually do. Probably both fed the young dove on the ground, although I witnessed only one feeding. Pigeons are not considered the most intelligent of birds, but this pair showed great adaptability in an exceptional situation and saved both of their nestlings.

The excellent care that the parents take of their eggs and young is not the least of the factors that have enabled White-tipped Doves to spread so widely over the tropics and subtropics of the western hemisphere.

3. The Birth of Beauty

Nature is not always friendly. By its vast carnage it distresses us; by its occasional outbursts of destructive violence it terrifies or destroys us; by stubbornly guarding its secrets from us it holds us aloof. But, in spite of all, its beauty binds us firmly to it, nowhere more compellingly than amid tropical splendor. Our planet owes its loveliness, in the large and in detail, above all to its mantle of green plants. Nevertheless, without animals plants would not, in the strictest sense, be beautiful. The explanation of this paradox should help us appreciate how much we owe to creative interactions between animals and plants.

Let us consider a flower that delights us with its beauty and fragrance. The flower has a certain symmetrical form which all but the most intransigent skeptic would admit exists extended in space independently of any observer. It reflects rays of light that consist of pulses or vibrations that differ in frequency and wavelength. It diffuses into the air certain molecules of complex structure.

This is all that physical analysis reveals of the flower's claim to loveliness and fragrance. I admire the blossom. Impinging upon my eyes, a small fraction of the luminous rays that it reflects is focused by each lens upon the corresponding retina in a pattern that reproduces, in miniature, that of the flower itself, much as light is focused by a camera on the sensitive film. Transmitted almost instantaneously by the optic nerves to my brain, excitations that originate in the many separate rods and cones of two separate retinas are, by a marvelous transformation that we are far from understanding, united in a single, delicately colored image. By another transformation that seems less complex, the molecules that diffuse from the flower to my nostrils are perceived as a delicious fragrance. But unless I am attentive and in the proper mood, I may fail to be pleasantly excited by the form, color, and odor that I perceive. An aesthetic response depends upon subtle psychic factors.

From this analysis, it appears that the

flower becomes beautiful at the moment when it is perceived by a sensitive observer. Its beauty does not reside in the flower alone, nor yet in the observer alone, but is created by their interaction. Every perception of beauty is a fresh creation, born of the fertile union of an appropriate object and a properly equipped observer. To this creative synthesis, the latter appears to make the larger contribution. A flowering plant, or the most elaborate work of art, is a structure far simpler than the manifold of eyes, nerves, and brain by means of which we perceive and respond to it. It is hardly an exaggeration to say that we find nature beautiful because we, as parts of nature, clothe it in beauty.

It is evident that to understand why nature contains so much loveliness we must pursue two lines of inquiry. On the one hand we must ask how it acquired the forms and colors that make it beautiful, while on the other hand we must ask how we happened to develop the psychophysical organization that makes us responsive to beauty. However, it will soon appear that these are not two independent lines of inquiry, for some of the forms that most embellish the natural world arose along with the capacity to perceive and respond to them by a process of reciprocal enhancement.

If we could have walked over the Earth in the Carboniferous period about 350 million years ago, we might have found it gloomy and oppressive yet not without much to admire. Probably on the best days white clouds adorned a sky as blue as that which covers us; but on the Earth we would have found little color except the prevailing green of forests dominated by imposingly tall, stiffly straight lepidodendrons, sigillarias, and calamites. Beneath these seedless trees we might have admired the intricately divided fronds of primitive ferns and seed ferns, but we would have searched in vain for bright flowers in the treetops or the dim undergrowth—they were still far in the future. Not a single song would have broken the gloomy silence of these dark forests, where the loudest notes were probably the croaks and roars of amphibians, some of which were huge. Insects, including dragonflies with a wing span of over 2 feet (61 centimeters), already flitted through the forest aisles, ate the tissues of plants, or preyed upon other insects. Regrettably, fossils fail to reveal the sounds and colors of that long-past age, but I surmise that the colors of these primitive creatures were mostly as drab as their voices were unmelodious.

Three hundred million years later, after the rise and extinction of the great dinosaurs, Earth in the early Cenozoic era was greatly transformed, chiefly by the replacement on a large scale of the old cryptogamic flora by flowering plants and the firm establishment of a new relationship between plants and animals. Undoubtedly, many of the earliest terrestrial animals ate plants, their only source of nourishment unless they preyed upon other animals that cropped the vegetation. But it is doubtful whether these primitive animals were of much service to plants, whose spores were carried by wind or water, as with modern mosses, ferns, and fern allies. The earliest flowering plants were apparently wind-pollinated, like modern oaks and poplars, and their seeds were scattered mainly by gravity or breezes. But as the flowering plants diversified, they depended increasingly upon animals to transfer their pollen and disseminate their seeds. The more varied the flora, the more necessary the services of winged pollinators became. In temperate-zone woodlands, composed of many individuals of few kinds of trees that in early spring are leafless or covered with thin needles like pines, wind pollination is effective but wasteful. In tropical evergreen forests composed of a great diversity of trees all mixed together, so that individuals of one species may be separated by many other species in full leafage, wind pollination is so inefficient that it is rare. At least in the New World tropics, more trees depend upon wind to scatter winged or plumed seeds than to transfer their pollen, which is chiefly done by animals.

With the exception of the relatively few species of plants that surreptitiously attach sticky or prickly fruits to fur or clothes, plants reward the animals that serve them. Flowers recompense pollinating insects, birds, and bats with nectar or nutritious excess pollen; fruits nourish the animals of many kinds, above all birds, that distribute their seeds. But, as businessmen well know, to offer goods of high quality is not enough, especially in a competitive market; to advertise them is essential. As ever more plants began to compete for the services of pollinators and seed-carriers, they developed means to attract appropriate animals to their flowers or fruits. One method was to diffuse an enticing odor from the open blossoms or ripe fruits. Another was to give them a color that contrasted with the prevailing green of woods and fields so that they might be more readily noticed. To plants, expert chemists that they are, the production of brilliant colors is no great problem. Dying leaves in northern lands, and to a minor degree in the tropics, often acquire attractive tints as though by accident, since these colors are apparently without value to the trees. As soon as bright hues proved serviceable to plants, they were certain to be ever more widely produced in flowers and fruits by the usual method of variation and natural selection.

We do not know if any of the pollinating or seed-dispersing animals could distinguish colors at the period when the first flowers or fruits acquired colors that contrasted with the surrounding verdure. Since different colors reflect different intensities of light, possibly these animals saw them as grays of varying shades. However, since at least some of the fishes, the oldest class of vertebrates, have color vision at the present time, it is not improbable that some fruit-eating terrestrial vertebrates could distinguish colors as soon as fruits became red, yellow, blue, or purple as well as green. The same might be true of pollinating insects and flowers. However, the widespread association of nectar, pollen, or edible fruits with diverse colors would promote the development of color vision in animals that lacked it or its improvement if some vestige of it were already present. As the verdure of woods and meadows began to be stippled here and there with the bright colors of blossoms and ripening fruits, animals became able to distinguish them, although not always as we do. Bees, for example, are sensitive to ultraviolet light, to which we are blind, and some flowers that we see as uniformly colored display a pattern to them.

Since color contributes so much to the beauty of nature, we must deplore the fact that the fossil record, especially of the small, usually delicate creatures that are often the most colorful, is so imperfect and helps little to trace color's increase in the living world. Nevertheless, I am convinced that the exchange of benefits between plants and animals contributed greatly to its increase in the animal no less than the vegetable kingdom. Animals that became highly sensitive to color, in the first place because it helped them find nectar or fruits, would themselves tend to become brightly colored because they preferred more colorful mates, as contemplated by Darwin's theory of sexual selection, or because distinctive color patterns helped preserve specific distinctness by preventing related species from mismating.

The fruit eaters and nectar drinkers include a disproportionately large share of the loveliest creatures on land. The most beautiful of insects, the butterflies, are nectar-sippers in the adult stage.

Many fruit- or pollen-eating beetles have brightly glittering shards. Bees tend to be too small to bear much bright color; but some of the furry bumblebees are colorful, and many of the euglossids, important pollinators of orchids, are shining metallic green or blue. The loveliest small and middle-sized birds are predominantly fruit eaters or nectar drinkers, including tanagers, honeycreepers, hummingbirds, and some of the cotingas in the Americas; sunbirds and barbets of Africa and Asia; birds of paradise in New Guinea; the fruit pigeons of Indonesia; and the cosmopolitan parrots, especially the flower-visiting lories and lorikeets of Australia, New Guinea, and neighboring islands. The most refulgent of the trogons, the Resplendent Quetzal, is one of the most frugivorous members of its elegant family. I would be sorry to leave the impression that fruit eaters and nectar drinkers include all of the world's most colorful birds. Many of the kingfishers and insectivorous bee-eaters and jacamars, to name only a few, are outstandingly beautiful. But, on the whole, the largely insectivorous species are less colorful than those that depend chiefly on fruits and flowers, and the raptors are consistently drab.

Colors have the most varied functions On flowers and fruits they serve to attract pollinators and seed disseminators, and on many animals they appear to attract mates or at least to reveal specific identity. On yet other creatures, including many insects, they have the contrary function of repelling predators by warning them that the bearer is distasteful or has effective means of defending itself. But warning coloration, other than bold patterns of black and white as in skunks, would be of little worth in the absence of animals with color vision, which among terrestrial creatures has been promoted by the interactions of animals and plants.

Beauty is born of the interaction of a responsive mind with an appropriate external object. It is delighted awareness of something that shares the Earth with us. No matter how many shapely forms and bright colors adorned its surface, a planet devoid of responsive viewers could not properly be said to contain beauty; it would have only the potentiality of beauty. Without creatures endowed with aesthetic appreciation, such a planet would lack something that seems to be essential for its own completeness. Our most convincing explanation of the adornments of the higher animals is the Darwinian theory of sexual selection, which implies something difficult to distinguish from an aesthetic sense. Melodious songs and the tastefully decorated gardens of the bower birds of New Guinea and Australia provide our strongest evidence that birds are not wholly devoid of such a sense. Nevertheless, only one animal, of the myriad that inhabit our planet, appears to have aesthetic appreciation so highly developed that he sees beauty in the most diverse objects and situations. Only man reproduces lovely forms in works of art, celebrates them in verse and prose, and photographs them. To understand how he acquired these capacities, we must return to the creative interactions of plants and animals.

The primates, the great branch of the mammalian class to which man belongs, first appear in the geological record about 70 million years ago in the late Cretaceous period, when flowering plants were replacing the older gymnosperms on a vast scale. Primates did not become abundant until the Eocene, when they spread over large areas of both hemispheres. Their evolution is intimately connected with that of flowering plants. With few exceptions, the more advanced primates are (and for long ages have been) diurnal, arboreal animals that roam the woodlands in family groups or larger clans, subsisting upon fruits, foliage, and flowers, often with an admixture of insects, birds' eggs, and such small vertebrates as they can catch. Even the most terrestrial of the primates, including baboons, gorillas, and men who have not lost their agility, can climb well, not by clinging to trunks and branches with sharp claws, like squirrels and most other arboreal mammals, but by seizing them with grasping hands. No feature of man's remote ancestors was more pregnant with promise than their grasping digits equipped with flat nails rather than claws.

To judge correctly the distance of their often tremendous leaps between the boughs of trees, monkeys need full binocular, stereoscopic vision, which is best achieved by eyes directed straight forward in a flat face, rather than situated toward the sides of the head, as in many vertebrates. To distinguish fruits and flowers amid green foliage, they have color vision, a precious endowment apparently not widespread among mammals, a large proportion of which are nocturnal and probably see things as varying shades of gray. Animals with color vision tend to be themselves brightly colored, as is most obvious among birds, although this tendency may be overruled by the need to remain inconspicuous to predators. Perhaps because they have long been snatched from treetops by the most powerful eagles, monkeys cannot afford to be too brightly attired. Moreover, hair seems not to lend itself to brilliant coloration as well as feathers do. Probably for these reasons, monkeys are never as brilliant as the smaller, more swiftly mobile birds that share the treetops and fruits with them. Although many primates are dingy animals, some monkeys have heads that are elegantly adorned with crests and whiskers of contrasting shades, as Darwin showed in the eighteenth chapter of *The Descent of Man and Selection in Relation to Sex*. The most vivid colors of primates are not on fur but on patches of scarlet, orange, blue, or violet bare skin, which are not always situated where human modesty would have them. These startling colors are evidently related to the primate's color vision, for they are rare on mammals that lack such vision.

When, probably in an era of decreasing rainfall and shrinking forests, the primates who were destined to evolve into man became adapted to a terrestrial life, they had acquired during their long previous history in trees two features that contribute greatly to appreciation of visual beauty: color vision and forwardly directed eyes that can concentrate upon whatever object engages attention. What was still needed to make these ancestors more fully aware of beauty was a more active mind, widely alert to everything lovely in the surrounding world rather than narrowly concerned with the vital needs of food, safety, and reproduction. And for this, in turn, a larger brain was apparently indispensable.

The clasping hands that these primates brought down from the treetops were, above all, responsible for the evolution of bigger brains. To these supple hands we owe the fact that of all the multitude of terrestrial animals, herbivorous, carnivorous, or omnivorous, man alone has acquired high intelligence, complex language, and the most catholic aesthetic sense in the animal kingdom. Without hands or some equivalent executive organ, high intelligence has so little survival value that it is not likely to be promoted by evolution; on the contrary, to know what needs to be done without being able to do it, to have brilliant ideas without the remotest possibility of realizing them because one lacks the means to manipulate objects, can be overwhelmingly frustrating. But when man's prehuman ancestors, the hominids, began to use their hands to fashion crude tools, intelligence became an important factor in evolution. Every increase in mental power, which implies brains that are bigger or of higher quality, made the hands that they directed more powerful instruments for promot-

Gibbon

ing survival. Every improvement in the structure of the hands, enabling them to perform more intricate operations, gave greater value to the guiding intelligence. Hand and brain evolved together by a process of reciprocal enhancement.

To this alliance of mind and hand we owe the growth of language. The calls and nonvocal means of communication of animals, together with their innate abilities, are adequate for their needs. Even if they could conceive large cooperative undertakings, their means for realizing them are, on the whole, woefully inadequate. But when the dawning human mind conceived projects conducive to a group's prosperity that could be carried out only by the cooperation of many hands, directed by the most intelligent or most creative member of the group, language became a powerful aid to survival. Mind, hands, and speech are so intimately linked in evolution that none would attain a high level of perfection without the other two.

Through many millions of years of slow evolution, first in trees and then on the ground, man acquired the forwardly directed eyes, the color vision, and the mind to appreciate beauty as no other animal appears able to do, along with the hands that enable him to create beautiful objects and the language that enables him to express his appreciation. For a long age after man acquired his present form and probably also mental ability, the harsh conditions of his existence, his constant preoccupation with getting food and escaping perils, prevented the flowering of his aesthetic nature. Nevertheless, by the late Paleolithic period we find Cro-Magnon man attentive to the grace of the animals that he hunted and reproducing his impressions of them in vivid drawings on the walls of the caves where he took shelter from a severe climate. By Neolithic times, people almost everywhere were expressing their love of beauty by decorating their pottery, fabrics, weapons, and other artifacts. Finally, in Classical Greece, the visual arts attained perfection that has never been surpassed and rarely equaled.

Man owes his abilities almost equally to the long arboreal stage of his evolution and a shorter terrestrial stage. Without both stages he would never have become what he is. Although the terrestrial stage was indispensable for the full development of his manual dexterity and associated intelligence, it has deplorable aspects. It transformed what was probably a fairly innocuous and peaceful, largely vegetarian arboreal primate, like most contemporary primates, into the most destructive predator that Earth has borne, one that has not only exterminated whole species of animals but has turned his hunting weapons against his own kind, making these weapons ever more powerful and deadly until he now has the means to accomplish his own extinction. Along with predatory habits, the terrestrial stage of man's evolution has fostered corresponding emotions and impulses, making him an exceptionally aggressive animal, capable of hatred, rage, sustained anger, and violence as few other creatures are.

Among the most precious and broadly significant attributes that man has acquired by means of a long and chequered evolution is his aesthetic sensibility, his capacity to see beauty and sublimity in the planet that he inhabits and the heavens that encompass it. This is what the cosmos needed for its fulfillment. For an immense age, Earth was covering itself with shapely and colorful forms, in the vegetable world and among the animals that plants support. White clouds adorned blue skies, rainbows arched across them, streams sparkled and foamed in sunshine, and, after the sunset glow faded, myriad stars jeweled the darkening heavens. But all this, and much more, would be wasted without appreciative spectators. A planet covered with such immense potentialities for beauty needs sensitive minds to make this beauty actual. An exceptionally favored planet, so extremely hospitable to life, would be incomplete without inhabitants grateful for the privilege of dwelling upon it.

Man is well fitted to fill the role of the appreciator, the grateful beneficiary, of all Earth's beauty and bounty. His aesthetic adaptation to his planet is wonderfully wide, far more perfect than his physiological adaptation. On land and on sea, in burning deserts and frozen polar wastes, in steamy tropical forests and chilly alpine meadows, in a thousand situations where he could not long survive without creating an artificial environment for himself, he so delights in beauty and wonder that he returns again and again, often at great expense and despite all discomforts. Integral parts of the planet that bears us, composed of its substance, supported by its circulations, we might regard ourselves as organs that, with immense travail, it has created for the appreciation of its beauty and sublimity.

The ancient Greek philosophers' quest of the supreme good was, in effect, a search for the proper end of human life, the activity that would most adequately give it significance and a sense of fulfillment. Perhaps Aristotle had the clearest insight when he declared that the supreme good is contemplation. Possibly he meant contemplation of eternal truths and abstract laws of thought, in the manner of his God, the Unmoved Mover, who apparently gave no attention to the concrete details of a universe that he moved by the attraction of his supreme perfection. But this is not certain, for Aristotle was also a naturalist who did not disdain the small details of nature as unworthy of his notice.

In any case, what we most need is more intense contemplation, greater enjoyment and appreciation of the beautiful natural world that surrounds and supports us. This should be not a languid aestheticism but a more vigorously intellectual interest that prompts us to learn about the things that attract us by their beauty, to share our discoveries with others, and, not least, to cherish and protect that which delights us. Although not new in the world, this attitude has been growing in recent generations. Nevertheless, it remains too rare in the great masses of men who live absorbed in the narrow human world, too often in its more violent and sordid aspects, unmindful of the wonder and mystery of the vast natural world whose resources they recklessly squander while they struggle for wealth, status, and superfluous luxuries. I believe that our salvation depends upon widespread recognition that nothing is more fitting and rewarding than appreciative contemplation of the splendor and beauty that surround us and the exaltation of our lives by this vision. By this course we can fill a niche in the natural world that is not adequately occupied by any other creature, give our lives high significance, and help fulfill evolution by the grateful enjoyment of its countless creations that need an appreciative mind to realize their beauty.

We have briefly traced the evolution of man's aesthetic endowment through the arboreal and then the terrestrial stage of the human lineage, but how can we explain it by current evolutionary theory? The evolutionist has no difficulty accounting for any aspect of structure or behavior that promotes the survival and multiplication of organisms. Awareness of the beauty that surrounds it might fortify an animal's will to live and reproduce. We might imagine that if the blue of the sky above us and the green of the earth around us were as oppressive as the dun sky and drab tenements of some old-fashioned industrial town, we would pine away. But to be depressed by colors that are not themselves injurious to health or associated with harmful conditions, even to recognize ugliness, is itself a manifestation of aesthetic sensibility, its negative side being the price we pay for our responsiveness to beauty. Moreover, people who live amid squalor may multiply more rapidly than those who dwell in more pleasant surroundings. When we invoke the theory of sexual selection to account for the splendor of certain birds and other animals, we seem to imply that beauty enhances reproduction and racial survival; but, unhappily, this seems hardly to apply to man in the present age, however much it may in remote generations have contributed to the charm of women. The only valid motive for begetting children is the generous desire to give them the inestimable boon of a rewarding life in a beautiful and interesting world, but this motive is too rarely present.

Perhaps the evolutionist will be forced to conclude that aesthetic sensibility is the incidental result of a psychophysical constitution that evolved to serve the basic needs of animal life. To see forms sharply helps an animal recognize enemies, social companions, and sources of nourishment; contrasting colors make them more conspicuous. To detect the color that reveals an eagerly sought food might stir an anticipatory pleasure, but this would hardly make a wide range of colors, not associated with any practical advantage, delightful in themselves. But why the pleasure? An automaton can be made to seize an object of a certain form and color without, presumably, feeling any pleasure, and certain famous philosophers have taught that animals are living automata, devoid of feeling.

To view some of the most precious achievements of organic evolution and of the world process in all its range, including beauty and our enjoyment of it, as accidental results is most unsatisfactory. Such narrow interpretations spring from the incomplete data with which the scientist works. That existence has a psychic pole or aspect we are certain—indeed, the thorough skeptic would say our only certainty, for in denying it we affirm it. Nevertheless, we cannot, by any means presently available to science, prove its presence, much less assess its quality, anywhere in the Universe except in our individual selves; we infer its presence in other beings, especially those most like ourselves, from behavior and analogy. To be able to observe the psychic aspect of creatures, as we do their structures and functions, might profoundly change our interpretation of the world process. If we could do this, we might conclude that beauty, love, knowledge, and other high values, far from being accidental results of the world process, are the goals for which it has been striving, gropingly and with many miscarriages, to be sure, but so persistently through such an immensely long age that its accomplishments are already great. We live in a beautiful world because, for eons and especially in regions with a tropical climate, plants and animals have interacted to produce lovely forms and colors and sharpen our awareness of them.

4. Castlebuilders

In the lowlands of southern Mexico and northern Central America, one with ears attuned to such sounds often hears a frank, good-naturedly persistent *bet chu, bet bet bet bet bét chu* issuing from the depths of a streamside thicket or abandoned plantation lushly overgrown. So dense and impenetrable by man is the vine-entangled vegetation amid which the author of these appealing notes lurks that it may be long before the birdwatcher catches a glimpse of the little brownish bird with rufous breast and wings. Although retiring, it is not shy, and when foraging near the edge of the thicket it may permit you to come near enough to discern the finer details of its appearance: its sharp, wrenlike, black bill, its red eyes, the fine white streaks on its dusky throat, its brown tail feathers frayed and worn by constant passage through the dense vegetation where it hunts for insects and spiders. These tattered feathers have given the name spinetail to this bird and numerous relatives that live farther south; although acuminate in some species, they are not spine-tipped like the tail feathers of woodcreepers and other climbing birds. For reasons which will presently appear, castlebuilders seems a more appropriate name for them.

In size, solidity of construction, and complexity of form, the nest of the Rufous-breasted Castlebuilder is to the nests of most other small birds as the towering castle of a medieval baron was to the huts and humble cottages of the common folk, his vassals, which surrounded it. With the castlebuilders, building appears to have become an obsession. Their great, elaborate castle of sticks occupies their unremitting attention during four or five months each year. Unlike most other birds, they add to it while incubating their eggs and even while feeding their nestlings. They continue to build in the waning light of evening when other diurnal birds are seeking their roosts. So absorbed do they become in the task of building the nest that they carry sticks to it while a person stands quietly in full view, little more than an arm's length away. I have cut my way noisily through the bushy tangle in which they lived to within a yard of their nest, without interrupting their building. While most other birds clearly build their nests merely as receptacles for their eggs and young, with the castlebuilders the construction of the nest appears to have become an end in itself.

In the humid Caribbean lowlands of Guatemala and northwestern Honduras, Rufous-breasted Castlebuilders lay the foundations of their nests toward the end of March or in early April. Their site, often near a stream, is usually in a vine-draped bush or small tree from 6 to 12 feet (2 to 4 meters) above the ground, rarely as low as 5 or as high as 20 feet (1.5 or 6 meters). Their bulky structure requires the support of a stout, nearly horizontal limb with several lateral branches or, better, two or more branches that are parallel and close together. In default of these, a network of interlaced vines may be chosen.

The male and female castlebuilders, alike in size and plumage, cooperate closely in the task of building. Until the nest is far advanced, their material consists almost wholly of coarse sticks, which they gather on or near the ground. Some of these are as much as 16 or 17 inches (41 or 43 centimeters) long and nearly as thick as a lead pencil. The heavier ones weigh from 2 to 3.5 grams.

Such sticks are no slight burden for birds that measure rather less than 6 inches (15 centimeters) from the tip of the bill to the end of the tail, and they work hard to raise these twigs through entangled vegetation and place them in the growing nest. Sometimes a castlebuilder, trying to fly up to its nest with a particularly heavy stick, grasped near the middle in its slender bill, is borne again to earth by the weight of its load; but it persists until it gains a low branch in the tangle where the nest is situated. Then, by a series of short flights and hops from twig to twig, with much flitting of its wings to keep its balance, it gradually attains its goal, displaying considerable skill in maneuvering its unwieldy burden between close-set branches.

When it arrives at the nest, the bird deposits its burden and industriously pushes and tugs until this fits securely in the desired position. Often, however, the stick is knocked from its bill by some unnoticed obstruction or dropped because of sheer weight. Then the tireless little bird follows it to the ground and undertakes anew the task of raising it by a circuitous route to the nest. Likewise, it descends to bring back sticks that fall from the nest while it is arranging them there. In this the castlebuilder differs from its relative the Firewood-gatherer of Argentina who, Hudson (1920) tells us, rarely retrieves material that it has dropped on the way to its bulky nest, until at length enough fallen sticks to fill a barrow accumulate beneath the nest tree.

The castlebuilders' growing nest soon acquires the form of a bowl. At this stage it resembles, except for the absence of a lining, the structure which many a larger bird finds a satisfactory receptacle for its eggs and babies. But the castlebuilders have only begun their ambitious undertaking. As they continue to build, they extend the bowl's rim on the side by which they habitually approach until it forms a gangway over which they now carry their sticks. Elsewhere, the birds build up the wall, at the same time contracting it toward the center until the bowl is completely arched over with sticks. Simultaneously, they add sides to the gangway, then begin to cover it over at the point where it joins the body of the nest, forming a tube or tunnel. The nest at this stage is an oven-shaped structure entered through a spoutlike projection from the side. The castlebuilders have now devoted nearly two weeks to their undertaking and already have a nest much bigger and better enclosed than that of any migratory northern bird, including those up to several times their size, but their work is only half done. The tunnel is still only partly covered and must be roofed to the end.

To leave open the end of this tunnel, as does the Slaty Castlebuilder of southern Central America, would make the entrance to their stronghold too direct for the Rufous-breasted Castlebuilders. As a further safeguard, like the winding entrance to some old Spanish fortress, they close the end of the tunnel and build an entrance chimney above it. This is a circular pile, often a squat tower, of twigs that are usually thorny and not only shorter but always much finer than those elsewhere in the nest. Through the center of this turret the birds enter and leave their home. The entrance, and frequently the whole length of the tunnel, is surrounded by a broad platform of coarse sticks on which the castlebuilders can alight and walk around.

When first covered, the nest chamber still lacks a roof that will shed water. To supply this deficiency, the birds now devote much labor to bringing short, thick sticks, broad pieces of bark, weed stalks, dry petioles of the Cecropia tree, and fragments of the broad leaf bases of giant herbaceous plants—all material quite different from that which forms the body of the nest—and heaping them above the chamber to form an efficient thatch that will keep the interior dry. This material is sometimes piled to a height of 8 or 10 inches (20 or 25 centimeters) above the chamber's ceiling.

Almost a month may elapse before the wonderful structure appears to be completed; but its exacting builders are not yet satisfied and continue to lavish their energies upon it until the eggs are laid, which, at nests begun early in the breeding season, may not occur until five or six weeks after the first sticks were laid in place. Even while they attend eggs and nestlings, the castlebuilders continue to bring material to their edifice, increasing its bulk without essentially altering its form until it becomes a monument to the tireless industry of the diminutive builders, out of all proportion to their own size. The final dimensions of the nest depend, among other things, on the time it has been in use and the nature of the site. If the supporting branches form too narrow a foundation, there comes a day when much of the material brought by the birds falls from the sides of the nest, which then fails to grow as bulky as it might have become if the birds had chosen a more ample support. Thus, there is considerable variation in the size of these structures. A fairly typical nest measured 29 inches (74 centimeters) in greatest length by 19 inches (48 centimeters) in height at the large end, where the nest chamber was situated. The interior of this chamber was from 5 to 6 inches (13 to 15 centimeters) in diameter, and the height from floor to ceiling 4.75 inches (12 centimeters). Above the chamber, the thatch was piled to a height of ten inches (25 centimeters). The tunnel that led into the chamber was 14 inches long and 1.5 inches in internal diameter (36 by 4 centimeters) and was surrounded by a platform 18 inches long by 17 inches wide (46 by 43 centimeters).

In May, the female lays, at intervals of two days, three or less frequently four eggs, which may be either pure white, white faintly tinged with blue, or a beautiful pale blue. The eggs in the same nest seem always to be of the same color, and sets of blue eggs are less frequent than those of white eggs. When first laid, they rest upon the bare sticks of the nest chamber, or perhaps a few green leaves or shreds of snakeskin inadequately separate them from this hard floor. While incubation is in progress, fresh leaves are added until at length the eggs repose upon an ample, soft lining which becomes the bed of the nestlings.

As they shared the work of nest building, so the male and female divide the duty of incubation. They sit rather impatiently, rarely continuing on the eggs for more than three-quarters of an hour. If the mate does not come to replace its partner after this interval, the latter may leave the eggs unwarmed while it seeks food. Sometimes the incubating castlebuilder calls loudly in response to notes of its mate which reach it through the wall of the chamber. Often a castlebuilder returns to relieve its mate after it has been incubating for only ten or fifteen minutes. At times, without waiting to be replaced, the bird neglects the eggs to give attention to the nest itself.

Constant attention is needed to keep in good repair a household as large and complex as that of the castlebuilders, who do not stint the time they devote to it. To preserve the nest in shipshape condition they lavish a degree of care that suggests loving pride and is pleasant to behold. Before entering or after leaving the tunnel, they usually find time for a round of inspection and nearly always notice something that requires adjustment. A broad piece of bark has fallen from the roof to the entrance platform and must be replaced in the thatch where it belongs; sticks that are slipping down must be pulled up into secure positions. Fresh leaves are brought into the nest daily and added to the bed beneath the eggs. For this purpose the castlebuilders select soft, downy leaves, especially those of a species of *Solanum* frequent in the second-growth thickets where they dwell. Since old, withered leaves are not removed, the mat beneath the eggs becomes thicker.

Not only do the castlebuilders find time for routine maintenance, but they take time to add to their nest and even to provide luxuries. They reminded me of hard-working human couples who start housekeeping with a minimum of comfort and by frugality and industry gradually acquire the luxuries they desire while they raise their children. We have already noticed that they continue to bring sticks to the nest while they incubate and to a lesser degree while they attend nestlings. But it is their endless search for the cast-off skins of snakes and lizards which occupies much of the time that they can spare from incubating and foraging for food. They seem never to have enough of these valued exuviae; if the bird arriving for its turn at incubation does not bring a stick or a green leaf, it will probably have a fragment of reptile skin, which it may either stuff into a crevice on the outside of the nest or take inside. Some are carried into the nest chamber, but most are used to cover the bottom of the tunnel. One pair, whose abandoned nest I opened, had completely carpeted their front hall with reptile skin, from the doorway right up to the nursery.

Castlebuilders seem never content to leave these precious bits of skin resting long in one place. Frequently they shift them from one position to another on the outside of the nest, or they carry a piece from the exterior to the interior, or they bring out a fragment from within. When its spell of incubation is done, a castlebuilder often emerges from the entrance chimney with a shred of skin in its bill and deposits it on the outside before flying off to forage. The habit of bringing the exuviae of reptiles to the nest is shared by other species of castlebuilders, by the Great Crested Flycatcher and related species that breed in holes in trees, by a number of wrens that nest in cavities and crevices, and by numerous other kinds of birds; but few, I believe, search for this material as assiduously as Rufous-breasted Castlebuilders do. Although nest maintenance and the search for reptile skins are done chiefly by the member of the pair who happens not to be incubating, frequently the eggs are neglected for many minutes while both engage in these absorbing occupations.

While for five hours I watched a pair of castlebuilders incubate beside the Río Tela in northern Honduras, the partner, probably the female, who spent most time with the eggs brought green leaves nine times and one stick. The other member of the pair compensated for his smaller part in incubation by bringing more material: snakeskin, seven times; sticks, five times; green leaves, twice; pieces for the thatch, twice—total, sixteen contributions. This bird apparently found a treasure trove of snakeskin, for in midmorning he came to the nest four times in quick succession, each time bringing a large, limp shred.

Castlebuilders and other members of the ovenbird family are among the few birds who consistently and apparently intelligently repair their nests. The only way I could learn exactly what a nest contained was to separate the sticks at the back or side of the chamber and peep or feel inside. In order to follow the course of incubation and the development of the young, I had to open the same nest repeatedly in this manner. After each examination, I pressed together the sticks that I had parted and stuck additional pieces into the gap that always remained, but I was never able completely to obliterate evidence of my intrusion. Returning to their nest, the birds would soon notice the alteration and promptly fetch fine twigs to stuff into the chinks that I had left, continuing this until one could not tell where the wall had been opened. The spot where the breach had been made would occupy the birds' attention for several days; it seemed to trouble them even after it had been, to my eyes, perfectly mended. Often they stuffed green leaves and fragments of reptile skin, which ordinarily were placed elsewhere, between the twigs where I had made a hole.

One day I purposely left a fair-sized gap in the side wall of a nest that sheltered two nestlings. The moment I moved aside, one of the parents, who had been fidgeting around a short way off, flew up bearing a stick that it thrust into the breach. Then it continued with great industry to bring twig after twig to the hole, flying down again and again to recover some that had fallen beneath the nest instead of going off to a distance for them. Meanwhile, the mate brought an insect to the hungry nestlings. Their domestic organization was excellent; neither bird appeared the least flustered by the damage I had done to their nest and neither uttered a sound. Each knew perfectly, without mutual consultation, what the situation demanded. When dusk was deepening into night, I returned and found one of the castlebuilders still pushing little twigs into the nearly obliterated gap, while its mate busily arranged sticks on the platform.

As nearly as I could determine, without unduly disturbing the routine of the nest by too frequent visits of inspection, the period of incubation was seventeen or eighteen days. The nestlings, blind and helpless, have pink skins with dark gray down on the top of the head, back, and wings. The inside of their mouths is yellow. As these nestlings grow older, they call for food with low, trilling chirps. Both parents brood and bring them small spiders, winged insects, and green caterpillars. Even the necessity to provide for four hungry mouths does not cause the adults to neglect their nest; they still find time to keep it neat, repair damages, and bring more twigs. They make neither vocal protest nor feint of attack when their young are removed from the nest in their presence, doubtless because such demonstrations would be of little avail in the presence of an enemy powerful enough to tear open their castle. They continue calmly to attend their nestlings while a man watches, unconcealed, only a few feet away.

While attending their nests, castlebuilders voice a variety of calls. The most frequent is the *bet chu, bet bet bet bet bét chu*, delivered with rising inflection and the emphasis on the final *bet*. The final *chu* is shorter and softer. A longer utterance sounds like *krr-r krrr-r krr kr kr, wita wita wita wita wita wita wit*. The first phrase is given slowly and deliberately, the second with increasing rapidity. At times, especially when their fledglings appear to be in peril, the castlebuilders make a low rattling or knocking sound.

I surmised that the parent castlebuilders might use their elaborate nest as a dormitory. Without crowding the nestlings, they could find, in the chamber, in the entrance tunnel, or in various protected nooks about the exterior of the great mass retreats that seemed safer and snugger than any that the surrounding thickets offered. But after the nestlings became feathered and no longer required nocturnal brooding, both parents slept at a distance from the nest. When I discovered how often these nests are pillaged, the advantage of this arrangement became apparent to me. The early separation of young and parents during the night when the latter could do little to protect their offspring decreases the probability that a predator will destroy the whole family at once. If, as frequently happens, the nestlings are eaten, the parents may survive to rear another brood. Only nests more immune to predation than the castlebuilders', such as those of certain woodpeckers, toucans, barbets, and swallows, are used as dormitories by the parents while the young are growing up in them.

One might surmise that nestling castlebuilders, growing up in a castle so cunningly designed and strongly constructed, would escape perils that exact so large a toll of young birds of most other kinds. But let a bird place her eggs where she will, swinging in a woven pouch from the tip of a slender twig a hundred feet in the air, in a hole in the heart of a tree trunk, in a burrow underground, or in an elaborate stronghold of sticks entered by a long and narrow passage, she is never able to elude all the mouths that hunger for her progeny. The castlebuilders are not exempt from the heavy toll levied upon eggs and nestlings of all kinds; their nests are despoiled with surprising frequency by small creatures, snakes or perhaps mammals, slender enough to creep through the tunnel and remove eggs or young without making a gap in the wall. As at nests of other kinds, any interference by man, no matter how slight or well intentioned, seems to increase their liability to destruction. Of the many that I watched, only one produced a fledgling. By refraining from touching the nest for two weeks after the eggs hatched, I probably contributed to its success. When finally I separated the twigs at the back of the chamber to examine the contents, a parent hopped very near me and made a low, rattling sound. As I pushed in my hand to feel for the nestling, it apparently fled down the tunnel and escaped through the doorway. Then, fluttering to the ground, it hopped away through short grass so rapidly that I succeeded in capturing it only after a chase of about 20 feet (6 meters). Fourteen or fifteen days old, it was already well feathered. It differed from its parents chiefly in its tawny rather than rufous-chestnut breast, its brown instead of red eyes, and its largely yellow instead of black bill. It disdained a small black berry offered by a parent, who then swallowed the fruit. Probably, if undisturbed, it would have remained in the nest a few days longer.

The structures that the castlebuilders erect with so much labor serve only for the one or two broods that they rear in a single season. After abandonment by their builders, the nests are often occupied by large black ants which swarm out when the supporting branch is shaken. Sometimes, indeed, the ants invade nests still used by the birds, for once I found a pair of castlebuilders carrying sticks to a structure that they had begun about a month earlier, and when I examined it I was stung by the ants that rushed out. If the nests remain free of ants, termites may invade them and build a network of galleries over their surface. Even if a nest happens to escape all insect invaders, during the long wet season it is so weakened by decay that it is unfit for use in the following year. The industrious castlebuilders must begin their nests anew each spring.

From the Caribbean coast of Honduras southward through the rainier parts of Central America and through Colombia to western Equador, the Rufous-breasted Castlebuilder is replaced by the Slaty Castlebuilder. About 6 inches (15 centimeters) long, it has dark grayish brown upperparts and slaty underparts, with bright chestnut-rufous on the crown and hindhead and much of the wings. Its deep gray throat is streaked with white, and a darker patch separates it from the chest. Its rather long brown tail is nearly always frayed by constant passage through dense vegetation. It has dull red eyes, a short, slender black bill, and legs and toes so shiny that their color changes from jet black to blue-gray with variations in the light in which they are viewed. The male and female are alike.

Avoiding closed woodland, Slaty Castlebuilders thrive in resting fields so densely covered with bushes and vines that a man can hardly pass through them without cutting a path with his machete; in neglected pastures where a maze of cowpaths intersect tangled growth; in hedgerows and streamside thickets. From birds foraging unseen in such dense vegetation I have heard answering throaty rattles so frequently that I feel sure that pairs remain intact throughout the year. In an old maize field in a forest clearing, thickly overgrown with molasses grass and pokeberry bushes, I often heard the castlebuilders as they became active at dawn. The first to awake called *churrrr* and was promptly answered with similar notes by its partner who, to judge by voice, roosted a short distance away. In the evening, I sometimes heard the pair communicating as they went to rest. My failure to find a dormitory strengthened my conviction that, like Rufous-breasted Castlebuilders, these birds sleep in dense growths of grass or bushes instead of in nests.

On the hilltop behind our house, one morning in July, I heard the rattles of Slaty Castlebuilders. Presently one perched in an orange tree with its head bent down and inward until its bill almost touched its breast and the feathers of its head and neck all stood out. Acceding to this invitation to preen, its mate perched beside it to nibble the outfluffed plumage. After performing this service for a while, the second castlebuilder picked up a twig from a small pile in a nearby fork. The first bird continued to hold its head down and its feathers erected as though it desired more of the same attention, but its mate preferred to move sticks. This was the only time that I saw any member of the ovenbird family preen another; but they are retiring birds whose less frequent activities mostly pass unseen by us.

It is difficult to see castlebuilders forage except at times like the height of a severe dry season, when the loss of much foliage increases visibility in their dense thickets. Then for brief intervals they may be watched as they hop over the ground and through the lower branches of shrubs, examining curled dead leaves which they sometimes pick up to extract anything edible that hides in them. Insects larval and mature, spiders, and other small invertebrates appear to be their only foods.

Once, while I sat in a blind before a nest of the Orange-billed Nightingale-Thrush, I watched a castlebuilder foraging through the thicket. Reaching the nest while the thrush was absent, the castlebuilder pecked at the mossy side of

Slaty Castlebuilder at nest, entrance at left

the open cup; then it stood on the rim and looked down for a moment at the two brown-flecked blue eggs. Then, of a sudden, it drew back its head and brought its sharp bill down hard against one egg, piercing the shell. The damage done, it at once continued on its way through the thicket. On returning to her nest, the nightingale-thrush immediately noticed the perforation in the shell and appeared to sample the contents of the broken egg, for she moved her bill mincingly, as though drinking. She sat for nearly twenty minutes on the damaged egg along with the whole one, then rose up, carried the former away, and in a few minutes returned to resume incubation of the single remaining egg. Since the castlebuilder hardly tasted the egg, it obviously did not pierce the shell to devour its contents. If this was the same individual that later came very close to my blind, it was a young bird who possibly had never before seen an egg. Attracted by the shiny blue objects in the nest, it was perhaps moved to investigate them.

Amid the low, tangled vegetation that the castlebuilders frequent, but often beside a path or small opening in the midst of the thicket, at its edge, or in a bushy hedgerow between open fields, they build their bulky nests in shrubs or small trees with close-set branches or in those overgrown with vines that provide additional support. Sometimes they choose a tree standing in a clear space some yards from the dense vegetation that they prefer. Nests that I have found were rarely as low as 1.5 feet above the ground and sometimes as high as 15 feet (.46 and 4.6 meters).

Calling back and forth with a rattling *churr*, the male and female work together to build their nest with sticks that they find on or near the ground a short way off. Although I have found Slaty Castlebuilders somewhat less confiding than their Rufous-breasted cousins, they have worked while I sat unconcealed only two or three yards away. In two and a quarter hours, the members of one pair made thirty visits to their nest, each time bringing a single twig which was sometimes branched and usually between 2 and 6 inches, rarely more than 8 inches (5 to 15 or 20 centimeters) long. A few days later, when they were building more actively, this pair brought twelve contributions in thirty minutes. Much of their time was spent pushing and pulling each piece until it interlocked with the other sticks to form a compact fabric. These castlebuilders may build for three or four weeks before they lay their first egg in a structure that is not yet finished.

Like the Rufous-breasted Castlebuilder's nest, that of the Slaty Castlebuilder is a massive edifice of tightly interlaced twigs, containing a rounded chamber entered through a long, narrow tunnel projecting from one side and roofed with coarser materials piled above it to form a thick thatch. It differs in several details. The end of the tunnel is open, permitting the birds to enter directly instead of through a low tower above it. Instead of lining the nest with whole leaves laid loosely on the bottom of the chamber, the Slaty Castlebuilder bites or tears green leaves into irregular fragments, usually with jagged edges, and with much cobweb binds them together in a fabric that with care can be lifted out in one piece. For this lining, castlebuilders prefer the leaves, densely covered with long-rayed, stellate pubescence, of the Berenjena, a white-flowered, thorny shrub often abundant in the low thickets where they dwell, or of some related species of *Solanum* with furry foliage. Eggs may be laid in chambers with little or no lining, which is gradually added as incubation proceeds.

These castlebuilders do not carpet their hallway with exuviae of reptiles, as do their northern cousins, but devote more time to seeking cobweb for their nest cup. However, they do search for snakeskin, which they may lay beneath their eggs or stuff into their walls, especially where they have been opened. They repair gaps no less diligently than the Rufous-breasts. A typical nest was 14 inches (36 centimeters) high at the end that contained the chamber, which was 9 inches (23 centimeters) in external diameter. From the back of the chamber to the open end of the tunnel it measured 17 inches (43 centimeters). The tunnel extended out from the wall of the chamber for 8 inches (20 centimeters), but taking into account the thickness of the wall, the tubular passageway that the birds traversed when they entered or left the nest was 13 inches (33 centimeters) long.

A most unusual nest had two entrances, each the long, tubular hallway, on opposite sides of the same central chamber. From end to end, this nest measured 22 inches (56 centimeters) and its height was 12 inches (30 centimeters). It was unlined when I first saw it, and apparently it never contained eggs. I doubt that such a nest would be more liable to predation than one of the usual form with a single entrance, and it would permit the parent or feathered nestlings to escape through one tunnel from an enemy that entered through the other, instead of being trapped inside. Nests built late in the year are often smaller, with thinner walls, shorter entrance tubes, and, despite the heavy rains of this season, less thatch than those made for first broods.

In the Valley of El General, I have found freshly laid eggs in every month from January to September, except the usually dry month of March. Nests with eggs are about equally numerous in all months from February to August, again excluding March. Although sets of three eggs have been reported, each of twenty-three nests that I examined held only two pure white eggs, of which the second was laid before 7:30 A.M., two days after the first.

The first nest at which I watched Slaty Castlebuilders incubate was situated in a small tree on an islet in a mountain torrent. To be certain that I could distinguish the sexes, I tried to make one of them mark itself by brushing against a tuft of cotton, wrapped around the end of a twig and saturated with white paint, that I stuck into the entrance of their nest. Several times they slipped past it as they entered or left without acquiring a readily detectable spot. Becoming annoyed with the presence of this "paintbrush" at their doorway, they tugged at it until they pulled it loose and carried it away. After this had happened twice, I changed my procedure. Cutting a slender stick about 7 feet (2 meters) long, I made a "paintbrush" of its end which I stealthily raised to the mouth of the entrance tube while a bird was inside. When I shook the supporting branch, the castlebuilder emerged past the obstacle without acquiring a noticeable stain. Instead of flying away, it hopped and flitted around, approaching to within two or three inches of the intruding object in a manner that seemed half inquisitive, half belligerent. Several times it came so near the end of my stick that I tried by a quick movement to touch it with the paint-soaked cotton, but always it flitted aside just in time to avoid contact. Later, I noticed that one member of the pair had a white mark at the base of its black bill. I called this bird Spot, and since this was the one that incubated at night, it was probably the female. The other, whose tail was exceedingly frayed, I dubbed Tattertail.

In my nearly ten hours of watching at this nest, Tattertail, the supposed male, took nine sessions on the eggs, ranging from 4 to 44 minutes in length and averaging 27.7 minutes. Spot incubated eight times, for intervals varying from 18 to 44 minutes and averaging 31.1 minutes.

The eggs were left unattended for five periods, ranging from 2 to 27 minutes and totaling 50 minutes. The two parents together kept the nest occupied, presumably while they incubated the eggs, for 91 percent of the time, during which each was present for a total of 249 minutes. Often the partner arriving for its turn of incubation hopped over the outside of the nest, sometimes making its circuit again and again, crawling through a narrow passageway formed by twigs that projected from the wall of the chamber and touched the top of the entrance tube. Then it entered, and a few seconds later the partner who had been sitting emerged. At other times, the bird coming to take charge of the eggs entered directly, with no preliminary inspection of the nest's exterior. The changeover was usually made in silence, more rarely with a *churr*.

Nearly always the castlebuilder arriving to incubate brought either a piece of downy leaf for the lining or a weft of cobweb for binding these leaves together. Dry grass blades for the thatch and sticks for the structure in general were brought chiefly on special trips rather than when a bird came to replace its mate on the eggs. For days these birds continued to be preoccupied with the spot where I had opened their chamber to examine its contents, working fresh twiglets into the wall long after it appeared to me to be perfectly mended. On two evenings, they engaged most actively in tidying and repairing their nest after sunset, when the Gray-capped and Vermilion-crowned flycatchers who roosted on the islet had already congregated in the tops of the trees above them. They continued to be very busy with their housekeeping, going over and over the nest, pushing in a stick here and there, now and again flying off to fetch another bit of material, until the light became so dim that I could hardly follow their movements. At last, when it was nearly dark, Spot retired into the nest to warm the eggs while Tattertail flew away, probably to roost amid dense vegetation.

On another day, midafternoon was the time of most active housekeeping at this nest. Now Spot's chief activity was pulling up on the roof and tucking into place dry grass blades that were continually slipping down. On this day, the castlebuilders did not tidy their nest in the last light of day, probably because they had attended to this at an earlier hour. On yet another afternoon, the castlebuilders' activity took the form of ridding their nest of ants. After bringing a piece of material to the structure and going over it thoroughly to tuck loose sticks into their places, Spot continued intermittently for ten minutes to pick off, with her slender black bill, tiny brown ants that in great numbers were swarming up and down the supporting branch and over the nest. Whether she swallowed them or threw them aside I could not learn. Later, after the nestlings hatched, I watched Tattertail pluck minute objects, apparently more of the same ants, from the nest in much the same manner. The vigorous sideward toss he gave his head after seizing each ant left no doubt that he threw it aside instead of eating it. I have seen birds of a number of other species remove insects, mostly ants, from their nests; but sometimes they are unable to stem a massive invasion of ants, which overrun the nest and destroy the nestlings.

Years later, I watched another pair of castlebuilders incubate in a nest more favorably situated for observation than most. In over seventeen and a half hours, covering all parts of the day, I timed twenty-eight completed sessions by both sexes which ranged from 2 to 120 minutes and averaged 25.2 minutes. The eggs were unattended for twelve intervals, ranging from 1 to 48 minutes and averaging 13 minutes. Only one of these intervals of neglect was longer than 22 minutes, and most were considerably shorter. During the active period of 14.4 hours when changeovers occurred, the eggs were attended for 82 percent of the time. These castlebuilders ended their day early; at 4:11 on a rainy afternoon, the female began her long nocturnal session, which continued until 5:34 next morning. After being relieved by his mate at 4:11 P.M., the male disappeared until the following morning. During the active period he incubated much more constantly than she, taking the only sessions that continued for more than an hour, but she compensated by starting her nocturnal session early.

In fifteen hours of active attendance at their nest, this pair of Slaty Castlebuilders brought green leaves for the lining sixteen times, coarse pieces for the thatch four times, snakeskin twice, one stick, and unidentified objects six times, making twenty-nine contributions. This does not include fallen materials that they retrieved from the tangle of vines (never from the ground) directly beneath their structure and carried back to it—all in addition to their usual chores of housekeeping.

At one nest, both eggs hatched eighteen days after the last was laid. At the earliest nest that I have seen, eggs were laid on January 11 and 13, and the second nestling hatched between 7:15 A.M. and 1:30 P.M. on February 1, after an incubation period of no less than nineteen days and no more than nineteen and a quarter days. The nestlings resemble those of the Rufous-breasted Castlebuilder. When two or three days old, two nestlings were fed twenty times in four hours, at the rate of 2.5 feedings per hour for each of them. Both parents brought them small insects, larvae, and spiders, always a single item held conspicuously in their bills. Although I saw them remove no droppings, the nest remained perfectly clean, whence I inferred that they were swallowed by the parents. These nestlings were brooded fourteen times, for intervals ranging from 2 to 24 and averaging 9 minutes. A parent was in the nest with them for slightly more than half of the four hours of my watch.

Even with nestlings to feed and brood, these parents did not relax their attention to the nest itself. At least three times in the four hours they brought something to add to it: a large fragment of reptile skin, a piece of leaf for the thatch, and an unrecognized object. They also spent much time patrolling the outside of their edifice and replacing loose pieces, or just idly hopping over it and resting near it. Once one of them, after brooding two minutes, tidied the nest for sixteen minutes, then entered to brood for nine minutes more until its mate came with food. On another occasion, one patrolled the nest for seventeen minutes, and again one devoted six minutes to arranging materials on the outside of the structure.

Just as I was ending my watch, a parent emerged from the nest and flew to the ground, where it voiced very sharp monosyllables, similar to the usual call note but stronger and more insistent. The mate promptly came and called in the same manner. Soon I noticed a slender green snake, possibly 2 feet (61 centimeters) long, stretched out in a tangle of green vines near the ground. The snake slid deeper into the massed herbage, and the castlebuilders, still scolding sharply, hopped back and forth through the tangle where it lurked, coming very close to if not actually touching it. I could not see whether they pecked the snake. These green serpents are great nest robbers, and to save the nestling castlebuilders I removed it. Almost at once the parents resumed feeding their young.

The behavior of the castlebuilders toward the snake was different from that toward a gray crested lizard, nearly a foot in length, including its long tail, which for well over an hour clung motionless to an upright twig less than a yard from the nest. The castlebuilders gave no evidence of having noticed the lizard, although it could not have escaped their sharp eyes. I inferred from their behavior that they regarded the lizard as harmless, but I chased it away for greater safety. Likewise, the birds paid no attention to the shiny little lizards called Lucias even when they crawled over the nest.

When I returned ten days later, hoping to watch these castlebuilders attend their feathered young, I found their nest empty, and they were already building a new structure about 40 feet (12 meters) away. Since whatever took the nestlings made no gap in the wall of the chamber, it must have entered through the tube. A number of other despoiled nests were likewise intact, but in other instances I noticed a larger or smaller gap, usually round, in the wall. From this, we may infer that nests are pillaged by two types of predators, one of which uses the same entrance as the castlebuilders themselves, while the other tears open the side of the nest. The former are probably snakes; the latter, mammals of varying size. If the predator begins to breach the wall, the parent has plenty of time to escape down the tunnel; but if the assailant creeps stealthily into this passageway, the parent may be trapped in its solidly built castle. This disaster might be avoided if the castlebuilders consistently constructed nests with two entrances, such as I once found.

One wonders whether parent castlebuilders are often trapped in their laboriously built structures. Yet it seems certain that this cannot happen frequently, for it would so penalize this type of construction that natural selection would eliminate it. At the same time, one wonders what advantage these bulky nests can have, what enemies they hold aloof from the eggs, nestlings, or brooding parents. In all closed nests, the parents appear to run the risk of being cornered, but most such nests are either high and relatively inaccessible, such as the holes of woodpeckers, or in vertical banks difficult to scale, like the burrows of kingfishers. The castlebuilders' stronghold is just as accessible as any open nest placed in bushes and thickets; it is far more conspicuous, by no means immune to predation, and impedes escape by the parents whenever the predator enters by the birds' own passageway. It is difficult to see what advantages may counterbalance these obvious disadvantages, unless it be that these birds enjoy building. However that may be, castlebuilders and their many relatives who are also great builders are among the most abundant birds in tropical America and temperate South America.

My efforts to learn how long the young remain in the nest have been frustrated not only by the frequency of predation but also by the habits of the birds themselves, for when I cautiously made a small opening in the nest to learn if they were still present, feathered nestlings fled down the entrance tube and escaped prematurely. Blocking their exit with a handkerchief did not help, for when I removed it they flew out instead of returning to their nursery. A sixteen-day-old fledgling that escaped in these circumstances flew about 20 feet (6 meters) and was too wary to be caught and replaced in the nest. This was in the afternoon; without my interference, the young castlebuilder would probably have remained at least until the next morning, so we may conservatively place the nestling period at seventeen days.

A month later I met, in the thicket near this nest, two young castlebuilders who were apparently its former tenants. They were confiding and permitted me to follow them closely along narrow cowpaths that wound through the low, dense thicket. Now, about forty-six days old, they hunted assiduously for themselves, investigating curled dead leaves in bushes and vine tangles up to 10 feet (3 meters) above the ground, and they were also fed occasionally by at least one parent. The young castlebuilders answered the adult's rattle with similar but weaker calls. The parent also repeated a sharp monosyllable that reminded me of a note of the Mourning Warbler but was higher in pitch.

These young castlebuilders, who appeared not to have begun to molt into adult plumage, resembled feathered nestlings in coloration. They lacked rufous-chestnut on crown and hindhead, and the rufous on their wings was paler and less extensive than on adults. The plumage of their bodies was much paler than on the parents, being generally olive above and grayish olive instead of slaty below. The young birds also lacked the dark, white-streaked throat of the adults. Their eyes were dark, not red as in the adults.

The third castlebuilder that I know well is the Pale-breasted, a bird slightly smaller and much paler than the two foregoing species. Both sexes have the forehead gray, the crown and hindhead cinnamon-rufous, and the remaining upperparts, including the tail and most of the wings, plain grayish brown, with a conspicuous patch of cinnamon-rufous on the wing coverts. The sides of the head and most of the underparts are gray of varying shades which fades to white on throat and abdomen and is streaked with white on the chest. Their bills may be wholly black or black with much gray on the lower mandible. Their eyes are yellowish brown, their legs and toes pale flesh color.

Widespread in South America from central Argentina to the Caribbean coast, this castlebuilder enters Central America only as far as the Térraba Valley in southern Costa Rica. Long established in the savannas around Buenos Aires de Osa in the middle of the valley, its upward spread was blocked by a great expanse of scarcely broken rain forest. With the destruction of this forest during the past four decades, it has extended its range to near the head of the valley, where I frequently meet it. Avoiding the dense, vine-entangled second growth that Slaty Castlebuilders prefer, the Pale-breasted inhabits lighter vegetation, including bushy neglected pastures, weedy fields, and expanses of coarse grass interspersed with low trees and clumps of shrubs. Occasionally these related species meet where their habitats overlap. The Pale-breasted Castlebuilder forages on or near the ground, hopping through tangled growth and beneath tussocks of grass, where it is difficult to see what it finds; probably, like other castlebuilders, it is almost exclusively insectivorous. It exposes itself more often than the Slaty Castlebuilder, not hesitating to fly 50 or 100 feet (15 or 30 meters) across a clear space, something its relative would do only in exceptional circumstances. Likewise, it often rises to a fairly high, exposed perch to deliver its notes.

The Pale-breasted Castlebuilder's sharp, emphatic *bet chu* reminded me of the opening notes of the Rufous-breasted Castlebuilder's longer phrase but seemed drier and harsher. This disyllabic phrase may be repeated incessantly for long intervals. I timed one Pale-breast who called tirelessly, at the rate of thirty-six to forty-one *bet-chu*'s per minute, for thirty-five successive minutes, during which it delivered about thirteen hundred of these phrases. I watched another castlebuilder who called for a long time while perching from 20 to 30 feet (6 to 9 meters) up in the open crown of a roadside tree, something I have never known the other two castlebuilders to do. With each repetition of the *bet chu*, the caller's throat feathers stood out momentarily, revealing their dusky bases and making a fugitive dark patch on its light throat. Another utterance is a low rattle or *churr*, often prolonged and readily distinguished from the shorter rattle of the Slaty Castlebuilder. A series of sharp, emphatic notes—*bip bip bip bip*—may be delivered alone or as a prelude to a long rattle. A more elaborate performance begins with a series of loud *bip*'s followed by a very prolonged rattle and terminated by *bet chu* several times repeated. The Pale-breasted Castlebuilder is a noisier, less retiring bird than the Slaty and the Rufous-breasted.

As in other castlebuilders, both sexes build. The few nests that I have seen at the northwestern extremity of the species' vast range were situated from 2 to 5 feet (.6 to 1.5 meters) up in a tussock of coarse grass, in a thorny shrub overgrown with molasses grass, or on a bushy roadside between open pastures. They had much the same form as the Slaty Castlebuilders' nests but were much slighter structures, composed of straws, pieces of herbaceous stems, and fine twigs, some of which were thorny. Few of these pieces were as much as 6 inches (15 centimeters) long; one measured 7 inches (18 centimeters) but was hardly 1/16 of an inch (1.6 millimeters) thick. The castlebuilders broke twiglets directly from shrubs and gathered others from the ground. All were carefully interlaced to form a firm, cohesive fabric through which little light passed. These nests were roofed with coarse grass blades and straws, some with roots attached. The bottom of the chamber was covered with fine, irregular fragments of downy leaves from a species of *Solanum* which were mostly added after the eggs had been laid on the bare floor, as with other castlebuilders. A feature of these nests that I never found at nests of these other species was a ring around the entrance composed of the bases of grasses or other small herbaceous plants. The short, stiff roots of these plants formed a bristly collar which may have discouraged the entry of certain small predators.

The upright part of one of these nests was 7 inches high by about 6 inches in diameter (18 by 15 centimeters). This contained a nearly spherical chamber about 4 inches (10 centimeters) in diameter, which was covered by a thatch 1 inch (2.5 centimeters) thick. The straight, horizontal entrance tube that led into the side of this chamber was about 6 inches (15 centimeters) long, so that the whole structure was about 12 inches in horizontal length, 7 inches high, and 6 inches in greatest width (30 by 18 by 15 centimeters). Another nest was 12 inches high, 11.5 inches long, and 6 inches wide at the chamber (30 by 29 by 15 centimeters). In other parts of the Pale-breasted Castlebuilders' range, they build bulkier nests. Some found in the Orinoco region of Venezuela by Cherrie (1916) were from 16 to 20 inches (41 to 51 centimeters) long, composed of dry, thorny twigs up to 5 inches (13 centimeters) long, all interlaced into a fabric hard to tear apart. In Trinidad, nests have been found from 2 to 30 feet (.6 to 9 meters) above the ground (ffrench 1973).

Although at nests of the Rufous-breasted and Slaty castlebuilders I could never see the eggs without opening the side of the chamber, at the small nests of the Pale-breasted Castlebuilder I could view them by pushing a little mirror, attached to a slender stick, down the entrance tube and applying one eye to the opening. Accordingly, I could learn with much less disturbance of the nest when the eggs were laid, which was early in the morning, the second two days after the first. Two to three is the usual number of eggs in the northern tropics, but at a nest in Venezuela I found four eggs, a number that is frequent beyond the tropics in

Uruguay and Argentina. These eggs are white or pale greenish, unmarked.

While they had only one egg, the pair at the roadside nest were exceedingly voluble. It was at this time that I heard one of them call *bet chu* for thirty-five minutes with hardly a pause. Part of this series of about thirteen hundred calls was delivered inside the nest. At intervals this bird's mate answered the *bet chu* with a long rattle. On other occasions, I heard the rattle issuing from the nest. They kept their nest tidy and brought more materials, including flattish scraps of wood, bits of bark, and twigs thicker than those in the walls, all of which added 3 inches (8 centimeters) to the thickness of their roof. They found bits of snakeskin and scraps of colorless cellophane to stuff into crevices about the nest.

After the third egg was laid, I watched these castlebuilders incubate for over twelve hours, covering all parts of the day. The male and female regularly alternated in the nest, but I could not always distinguish the sexes. Ten completed sessions of both of them ranged from 29 to 102 minutes and averaged 59.2 minutes. The eggs were unattended for only three intervals, lasting 27, 18, and 29 minutes. One or the other member of the pair was in the nest, presumably incubating, for 89 percent of the eleven hours not included in the nocturnal session. Now the castlebuilders were much more silent than before they started to incubate in earnest. Although they continued to keep their nest in order, they devoted less time to this than do castlebuilders such as the Rufous-breasted and Slaty with larger and more elaborate structures. During the twelve hours of my vigils, they brought only three pieces when they came to incubate; one was a green leaf and the others unrecognized. One of the partners carried a piece of snakeskin from the nest when it ended its turn on the eggs.

One morning while the castlebuilders were absent, a Southern House-Wren approached stealthily through the surrounding herbage, examined the exterior of the nest, and filled its bill with bits of snakeskin that had been tucked into the outside of the tunnel. Then it entered the nest with its plunder, but promptly emerged to fly away with the snakeskin, voicing a little *churr* as it went. As far as I could see in my mirror, the wren did not harm the only egg then present.

This nest was on a low bank beside an unpaved road, used by heavy trucks and cars with four-wheel drive. The incubating castlebuilders were not disturbed by the noisy passage of these vehicles, sometimes with loudly clanking tire chains, only four yards from where they sat in their snug chamber. My sojourn in this locality ended before their eggs hatched.

Castlebuilders belong to the genus *Synallaxis* which, with 37 species, is the largest and most widely distributed in the ovenbird family, the Furnariidae. One of these, the Rufous-breasted Castlebuilder, is found only in northern Central America and southern Mexico; two others, the Slaty and the Pale-breasted, are shared by Central and South America; and all the rest belong to the southern continent and closely adjacent islands. The 214 species of ovenbirds are confined to the western hemisphere, where they range from tropical rain forests to high, cold páramos and punas and over the south temperate zone to bleak Tierra del Fuego and the Falkland Islands. None reaches the United States. Lacking bright spectral colors, ovenbirds are clad in shades of brown, rufous, gray, and buff, sometimes with spots or patches of black or white in many attractive patterns. The sexes are always similar. In structure and habits they are so diverse that Vaurie (1980) could write, "The ovenbirds behave like and often resemble morphologically to an amazing degree almost any passerine family one can think of, provided it belongs to a group lacking or poorly represented in South America."

Certainly no other avian family has more varied nesting habits or exhibits such diverse architecture. The family is named for a genus of brown, largely terrestrial birds which over much of South America build with mud or wet clay a substantial nest with a domed top and doorway in the side—a miniature of the clay baking oven widely used in tropical America—on top of a fence post or in the stout crotch of a tree. Fittingly, these birds are called horneros, from *horno*, the Spanish word for oven. With one partial exception, all members of the family nest in closed spaces, such as holes in trees or walls or burrows in the ground, or build covered nests, more often of sticks or other vegetable materials than of clay. Among the most impressive of these are the structures of interlaced sticks which Rufous-fronted Thornbirds hang conspicuously from South American trees or vines. An inaccessible nest that I saw on the llanos of Venezuela was about 7 feet (2 meters) tall and appeared to contain eight or nine chambers. Nests 3 or 4 feet (1 or 1.2 meters) high, with four or five rooms, each with its separate entrance, are not uncommon (Skutch 1969a, 1969b, 1977).

5. Parental Care and Family Life in Birds and Mammals

Birds have many ways of nourishing their newborn young; mammals, only one. The differences in their methods of caring for their progeny during the first weeks or months after birth are responsible for far-reaching differences in the family life of the two great classes of warm-blooded vertebrates.

The birds around our Costa Rican home exemplify the two most widespread methods of feeding nestlings. Parent tanagers, wood-warblers, vireos, thrushes, wrens, larger finches, and most flycatchers approach their nests with fruits or insects visibly protruding from their bills. As the adults alight on the nest's rim or in the doorway of a covered structure, the nestlings raise or thrust forward mouths widely gaping to expose a bright red, orange, or yellow interior, bordered at the corners by white or yellow flanges of skin that make the mouth look bigger. Often their mouths appear to open like brilliant flowers. Probably these colors, no less than the cries of hungry nestlings, stimulate the parents to feed them; they certainly provide conspicuous targets that must be helpful in a dimly lighted hole or covered structure such as many tropical birds build. In a trice, the parent sticks the food into one or more of the mouths, then flies off for more or else settles down to brood the nestlings.

Very different is the way that hummingbirds, pigeons, seedeaters, grassquits, Bananaquits, and some woodpeckers feed their young. These birds fly to their nests with nothing visible in their bills or mouths. A parent hummingbird thrusts her long, sharp bill so far down the nestlings' throats that one watching the process for the first time fears for their lives. Then, with jerky movements that reveal muscular effort, she pumps aliment into her usual brood of two, commonly feeding them alternately. When hungry, a nestling pigeon rises up in front of the brooding or guarding parent, who takes its bill into the side of its mouth. While still young and blind, pigeons are usually fed singly. After their eyes open, the two nestlings that comprise the brood of many species manage to insert their bills into opposite sides of the parental mouth. Then their heads bob up and down with that of the parent who, with obvious effort, regurgitates nourishment to both of them.

Why these differences in the ways of feeding young often found in birds of the same family? They appear to be related to the nature of the nestlings' diet. It would be difficult for hummingbirds to hold in their slender bills much of the nectar from flowers, often mixed with tiny insects caught in the air or gleaned from foliage and bark, that they give to their young. Newly hatched pigeons are nourished with almost pure milk secreted by the adults' crops. As they grow older, this curdlike substance is mixed with increasing amounts of the seeds, fruits, or insects that the parents gather. Obviously, the milk produced in the pigeon's alimentary tract must be regurgitated, and the addition of much solid matter to the diets of older nestlings does not change the procedure. The small tropical finches known as seedeaters and grassquits bring their young many minute grass seeds, often mixed with small insects, that are most conveniently carried in the throat or deeper regions of the digestive tract. Among northern birds, seed-eating goldfinches, siskins, and crossbills also feed by regurgitation. Bananaquits, which are sometimes placed in the wood-warbler family and sometimes with the honeycreepers, have a diet much like that of hummingbirds, consisting of nectar and tiny insects that they regurgitate to their nestlings. Honeycreepers that eat much fruit and larger insects approach their nests with the items lined up in their long bills.

The situation among woodpeckers is particularly instructive. Those that raise their nestlings on fruits and insects of fair size bring these offerings conspicuously in their bills. The numerous species whose diet consists largely of ants, including flickers and Pileated Woodpeckers in North America and Green and Black woodpeckers in Europe, feed by regurgitation, as they could not carry enough of these insects in a narrow bill. An interesting exception is the Olivaceous Piculet, a pygmy woodpecker that is one of the smallest of birds. It flies to the nest cavity with its diminutive bill overflowing with larvae and pupae of ants, the nestlings' principal fare. For a piculet, even an ant pupa is a fairly large object. When versatile woodpeckers that do not regurgitate vary their nestlings' diet with liquid food, such as sap from a boring in the bark of a tree or syrup from a feeder, they fill their bills with insects, seeds, or bark, which they use to sop up the fluid and carry it to the nest.

Since a crop or stomach holds more than a bill, birds who feed their young by regurgitation need approach their nests much less frequently than those who feed directly, a saving in movement and energy. Moreover, fewer trips to the nest diminish the probability that the parents' activity will guide predators to it. I have watched woodpeckers that carry food in their bills give their nestlings more meals in an hour than some that regurgitate bring in a day to broods of the same size. Nestlings fed by regurgitation, including hummingbirds, Bananaquits, certain parrots, and small finches, have a swelling on the side of the neck caused by the expansion of the esophagus in which the food rapidly poured into them is stored until needed. For marine birds, including petrels, shearwaters, and albatrosses, who may forage hundreds of miles from the island where they nest, feeding by regurgitation is the only feasible method. These far-ranging parents deliver massive meals of more or less predigested food to their single nestlings, thereby enabling them to remain unfed for days together.

Birds of prey and freshwater birds have still other ways of feeding their young. I was amused by a Double-toothed Kite that nourished feathered nestlings largely with insects caught in the air. The parent would hold an insect beneath a foot while it tore off many tiny bits and passed them alternately to its young, reserving some for itself. However, when it brought a lizard, much larger than any of the insects but not so easy to dismember, it permitted a nestling to swallow it whole! Older raptorial nestlings often help themselves by tearing pieces from the dead prey that a parent deposits in their eyrie.

The predigested liquid soup that herons, egrets, and bitterns regurgitate to their newly hatched young is replaced by whole or partly digested fish, frogs, and other items as they grow older. The young heron or bittern grasps the parent's bill in a scissorslike grip and often wrestles strenuously with it until the food emerges. Quite different is the method of spoonbills and ibises, who hold their long bills open while the young pick regurgitated food from their throats. This is also the way of cormorants, boobies, pelicans, and related birds. To feed an older nestling, a pelican opens its great pouched beak and permits the youngster to stick its whole head inside while it rummages blindly in the capacious gullet for something to eat. The struggle may become violent and result in injuries to the parent or young, as sometimes happens to Pink-backed Pelicans in Africa.

In response to a chick's pecking at its bill, a parent gull regurgitates a mass of partly digested food on the ground, then picks it up piece by piece and passes it to the young gull, who may also feed itself from the supply before it. Most terns receive small fish or insects directly from the parent's bill, but the pelagic Sooty, or Widewake, Tern feeds by regurgitation in response to pecks on its bill, much in the manner of gulls.

All the foregoing birds are altricial or semialtricial, fed in the nest or near it by parents who bring food from a distance. Nidifugous birds, whose young leave the nest soon after they hatch and accompany their parents in quest of food, have different ways of feeding their progeny. Subprecocial chicks, including grebes, rails, cranes, guans, and some pheasants, take their meals directly from their parents' bills. Fully precocial chicks feed themselves from the beginning with items that the parent may uncover by scratching and point out to them, as the domestic hen does; or the young follow a parent while they find their own food, the method widespread among ducks and shorebirds.

Feeding nestlings is the part of parental care that makes greatest demands upon the adults' strength, and it is accordingly that in which both parents most frequently cooperate. If the male bird does anything more than defend the territory and the nest, he feeds his offspring. The next most frequent contribution of males to the reproductive effort is helping to build the nest; many male birds aid in nest construction as well as feeding but fail to incubate. However, in the majority of avian families below the songbirds (Oscines) in our systems of classification and in not a few songbirds, the male also shares incubation, and, as among woodpeckers and nonparasitic cuckoos, he may spend more hours on the nest than his mate does. Thus, in a large proportion of the world's birds, males help rear the progeny as far as they are physiologically able, which means that they do everything except lay the eggs. And often they help their mates form eggs by feeding them as the time for laying approaches.

To be sure, many exceptions occur in a class of vertebrates so large and varied as the birds. Among precocial birds, which are spared the strenuous effort of carrying many meals to the nest, males are most likely to leave all the work of raising a family to the females; although some of them, including male rheas, tinamous, jacanas, phalaropes, and certain other shorebirds, bear the whole burden. Among altricial birds males that remain aloof from nests are hummingbirds, manakins, most birds of paradise, and some cotingas, woodcreepers, and American flycatchers.

In strong contrast to the birds' many methods of nourishing their callow young, mammals have only one, familiar to everybody. The mammalian mother has only to eat well and make her nipples accessible to her babies of whatever kind by standing, sitting, or lying in an appropriate posture or, as among marsupials, bearing her young in a pouch or otherwise attached firmly to her. The mammalian father is as little able to share this occupation as he is to participate in the development of the embryo that grows within her; at most, he can aid indirectly by bringing food to the mother while she is pregnant or lactating, as a few male mammals do. The great advantage of viviparity over the oviparity far more widespread in the animal kingdom is that it makes the embryo or fetus as mobile as the parent who bears it; the mammalian mother fleeing an enemy too powerful to confront carries her unborn young with her and may thereby save them. The bird in similar circumstances usually abandons her eggs or callow young, with the result that a very large proportion of all eggs and nestlings are destroyed by predators of many kinds.

But why could not mammals gain the advantage of viviparity without, unfairly as it seems, placing the whole burden of rearing the young upon the mother? Why could not the father share their subsequent nurture by providing milk for his offspring, as male pigeons do? The answer appears to be that contact with the eggs while incubating—a necessary stimulus for the secretion of crop milk in each individual pigeon—was a prerequisite for the evolution of this capacity in both sexes. Lacking close association with the developing embryo, male mammals failed to evolve lacteal glands for feeding the young after birth. Largely, I believe, as a consequence of the different systems of parental care associated, respectively, with oviparity and viviparity, we find in birds and mammals contrasts in the patterns of family and social life no less striking than those between feathers and fur or between flying and walking.

The first great contrast that we notice is in the prevalence of monogamy in the two classes. Lack (1968) calculated that 93 percent of all nidicolous, or altricial, species of birds and 83 percent of nidifugous birds are monogamous. For the two categories together, monogamous species account for 92 percent of the whole number, the remainder being polygynous, polyandrous, or promiscuous. Included among the monogamous species are those in which a single male and female stay together while raising one brood, or for one breeding season, as well as those with more enduring pair bonds. Although among swans and geese, which form lifelong bonds, the mated male and female migrate together, for most smaller birds migration, or wide wandering during the inclement season, disrupts pairs. If both partners survive the many hazards of long journeys, they frequently reunite on their territory or nesting site of the preceding year. A large proportion of the permanently resident birds of mild climates live in pairs throughout the year and apparently mate for life. Birds have a strong tendency to remain permanently paired wherever ecological factors favor continuous residence.

I am aware of no comparable estimate of the incidence of monogamy among mammals, which on the whole are more difficult to watch than birds and have been less widely studied in their natural habitats, but I suspect that the percentages of monogamy and other forms of association of the sexes are approximately the reverse of those for birds, with possibly no more than 8 or 10 percent of the mammalian species exhibiting monogamy. Although we can point to whole families of birds that are predominantly if not wholly monogamous, monogamy is scattered thinly through the orders and families of mammals. Among the known examples of monogamy are a few of the many kinds of bats; beavers, certain deermice, and possibly some Old World porcupines among rodents; hooded seals; coyotes, jackals, foxes, European badgers, and apparently wolves among carnivores; the Dusky Titi and related species, gibbons, siamangs, and some humans among primates.

What is the reason for this great contrast in the prevalence of monogamy in birds and mammals? In a much discussed book, Wilson (1975) opined that the ecological conditions that seem to account for all known cases of monogamy are (1) territories with such a scarce and valuable resource that two adults are needed to defend them against other animals; (2) physical conditions so adverse that two adults are necessary to confront them; and (3) the great advantage of early breeding, which is facilitated when pairs are maintained. If all this be true, we must conclude that birds in general live in less productive territories than do mammals and are harder pressed to defend them, that they inhabit more inhospitable environments, or that early breeding is more advantageous for them than for mammals.

The reason why monogamy is so much more prevalent among birds than among mammals is undoubtedly the fact that the avian system of reproduction favors equal sharing by the two sexes of all parental offices, whereas the mammalian system does not. In all grazing and browsing animals and many others, the male can contribute little more than defense to the welfare of his progeny, and, if small and weak, his protection is of little worth.

It is no accident that among the Canidae (wolves, foxes, and their relatives) monogamy is, as far as we know, more prevalent than in any other mammalian family. Although the males cannot suckle the cubs, they can contribute substantially to raising them by bringing food in their stomachs and regurgitating it for the nursing mother and for cubs old enough to eat solid food—an ability rare, if not unique, in the whole mammalian class. Among beavers, the monogamous male plays a most important role in supporting his family.

Another consequence of the avian system of reproduction is that it enables immature or nonbreeding individuals to assist in rearing young, usually but by no means always later broods of their own parents. Juveniles of all three species of North American bluebirds have been watched bringing food to brothers and sisters only a few weeks younger than themselves; and similar precocious participation in parental activities has been reported in many other birds, including Barn Swallows in both Europe and America and House Martins in Europe. A long list of avian species, which increases yearly as more birds of the tropics and subtropics are carefully watched, practice cooperative breeding: nonbreeding individuals remain with their parents (or occasionally neighboring parents) through the second or even later years of their lives, helping to feed and defend the nestlings and fledglings, less often to build the nest, incubate, or brood the young.

In the United States, cooperative breeding is best developed in the Red-cockaded Woodpecker, Acorn Woodpecker, Florida Scrub Jay, and Gray-breasted, or Mexican, Jay. Elsewhere it has been found in Common Gallinules, Purple Gallinules; bee-eaters, hornbills, wood-hoopoes, and weaverbirds in Africa; Old World kingfishers; nunbirds, araçari toucans, jays, and wrens in tropical America; babblers in Asia; wood-swallows, wren-warblers (fairy wrens), honeyeaters, and mud-nest builders in Australia, to mention only a few of the many examples that have been studied. Understandably, the closest parallel among mammals to cooperative breeding in birds is found in the Canidae, for these quadrupeds are better able than most mammals to help parents raise their young by bringing food, as well as guarding them while the parents are absent. Among wolves, coyotes, jackals, and foxes, the parents may be helped by one to five nonbreeding individuals. In Golden Jackals and Silver-backed Jackals, the helpers are grown young from earlier litters who may remain with their parents for as long as two years. They groom and play with their younger siblings as well as feed them (Moehlman 1980).

Of all mammals, the industrious, much-persecuted beavers most resemble in their family life the birds that form the most enduring bonds with their offspring, and indeed they share many features with the ideal monogamous human family. The lodge of sticks and mud, which reminds us of the wattle-and-daub construction of primitive human huts, is built in the midst of a pond with an underwater entrance for safety. Here a monogamous pair dwell with their young until the latter are about two years old, much as certain woodpeckers, toucans, wrens, and other birds continue to sleep in a hole or nest with offspring who have long been full-grown. Although the father and his older children do not, like many birds, feed the callow young directly, they contribute substantially to the support of the litter by working together, with the senior male as leader, to construct and maintain the dam and canals that ensure a continuous supply of the bark upon which all the family, youngest as well as oldest, ultimately depend for nourishment. Here again monogamy and cooperative breeding are outgrowths of a pattern of reproduction which permits individuals other than the mother to contribute substantially to rearing the young.

It appears that the avian system is more compatible with the realization of some of the perennial, widely held ideals of mankind and the solution of certain of its most pressing problems than the system which, whether we like it or not, we share with the other mammals. Equality of the sexes would be much more readily achieved if, like birds, the human male were anatomically, physiologically, and psychologically equipped to share the duties of child raising almost equally with the female. Moreover, the sexual equality of birds frequently extends to appearance no less than to behavior. Although in northern lands, where to escape inclement weather so many of the birds undertake migrations that disrupt pair bonds, the most colorful male birds often have more plainly attired mates, among permanently resident and continuously mated tropical birds the brightly colored male and female are frequently so similar that it is difficult or impossible to distinguish them.

Anthropologists have contended that the human female's readiness for sexual intercourse throughout the estrous cycle, a situation rare or absent in other mammals, is an evolutionary adaptation for keeping a male attached to his wife during the long years when children need the care of both parents. Certainly, if one member of a couple is sexually active while the other member is not, the stage is set for promiscuity; but the whole history and literature of mankind, the harsh penalties for adultery in many societies, attest unmistakably to the dismal failure of continuous sexual readiness to safeguard monogamous fidelity. In this context, too, the avian system appears more effective. After the breeding season, the reproductive organs, especially the testes, shrink in size and become inactive. Nevertheless, throughout the long interval when no eggs are laid, the male and female of many species, now virtually sexless, remain closely associated, held together by shared territory, responsive singing or calling, mutual preening, and perhaps most of all by ties formed while they worked in closest harmony to raise one or more broods. The constantly paired birds of mild climates appear to exemplify platonic love to a degree difficult for humans to achieve.

Finally, the avian system makes casual and careless parenthood, the bane and shame of mankind, virtually impossible. To bring a nestling or chick into the world requires preparation and patience. The bird, with a mate or alone, must find an adequate nest site, in most species build a nest, and incubate faithfully for from eleven to eighty days before the chick escapes the shell—the equivalent of mammalian birth. Some of the Australasian megapodes, or mound birds, which do not incubate but hatch their eggs by heat of fermentation, solar heat, or a carefully regulated combination of both, work exceedingly hard to maintain the proper temperature of their incubation mounds. Even the cowbirds, parasitic cuckoos, and a few others that foist their eggs upon unwilling foster parents must watch carefully to drop eggs into their victims' nests at just the stage that will ensure proper incubation and care of the nestlings. If human reproduction required as much careful preparation as birds must undertake, the world's most urgent problem, the limitation of man's numbers, would be much more readily solved. Which makes me think it a pity that we are not hatched from eggs! To warm them with the heat of our bodies alternately with our spouses in the widespread avian pattern would give the opportunity for thoughtful repose that is sadly lacking in the too-feverish lives of many of us.

6. *A Charming Thief*

In March, the driest, warmest month of the year in southern Costa Rica, the Madera Negra trees planted as living fence posts around our garden shed their pinnate leaves and cover their long, slender branches with racemes of pink flowers. The rows of delicately tinted trees gladden our vision and help us endure the oppressive, smoky atmosphere of the last month of our short dry season, when farmers are clearing and planting hilly or rocky lands in anticipation of the approaching rains. The flowering trees attract many bumblebees; and wintering Baltimore Orioles add vivid patches of orange to the floral display as with sharp bills they probe the pealike blossoms. The few Orchard Orioles that I have seen in this elevated valley were visiting the Madera Negra flowers. From time to time an enchanting creature arrives unannounced to poise on invisible wings beside the long flower clusters.

Only a hummingbird could dart and hover with such exquisite mastery of the art of flight, but this bird's sylphlike form and long tail make it graceful beyond most members of a family renowned for delicate beauty. The immaculate snowy white of all the lower surface of its body, contrasting so prettily with the glittering green of its upper plumage, gives the Purple-crowned Fairy an elegance rare even among hummingbirds. If the charming visitor happens to be a male, his forehead and crown flash metallic violet when turned toward the watcher. The more numerous females have the forehead and crown green, like the back. Both sexes have black faces and short, straight, sharp-pointed bills.

These Purple-crowned Fairies seem to come from a distance to visit the flowering Madera Negra trees, for through much of the year these hummingbirds are scarce in this vicinity. Yet, strangely enough, the pink blossoms are largely ignored by the more abundant resident hummingbirds, including the Rufoustailed, the Charming, and the Snowybellied. Aside from March, the flowering time of the Madera Negra, I see the Fair-

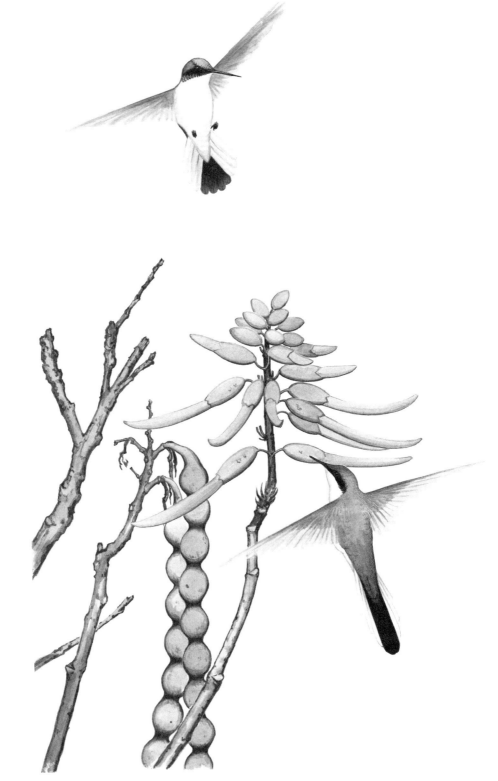

Purple-crowned Fairies at flowers and pods of the Red Poró, Erythrina berteroana

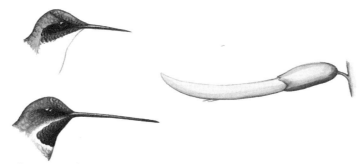

Purple-crowned Fairy (above) and Long-billed Starthroat at a poró flower

ies most frequently at the beginning of the drier weather in December and January, when the Red Poró trees in our garden shed their foliage and display their scarlet flowers on leafless boughs. Although the poró is, like the Madera Negra, a member of the bean family, its flowers have a very different appearance. All the petals are greatly reduced except the fleshy standard, which is long and slender and tightly folded lengthwise so that it resembles the blade of a sword or a machete. Its base is surrounded by the thick, tubular calyx, which encloses the rudiments of the other petals. The compact structure of the three-inch-long flowers makes their nectar inaccessible to nearly all of the local hummingbirds; of the numerous kinds that frequent our garden, only those with the longest straight bills and those with the shortest bills take an interest in them.

When the poró trees begin to blossom as the wet season ends, a Long-billed Starthroat Hummingbird with a flashing magenta gorget takes possession of them and expels all intruders. Plainly attired Scaly-breasted Hummingbirds, abundant in the vicinity, suck the sweet fluid from the scarlet flowers when the larger Starthroats are not watching. Both of these hummingbirds have bills and tongues long enough to reach the nectar by inserting their beaks into the slit along the lower edge of the flower's folded standard. But the shortness of the Purple-crowned Fairy's bill is compensated by its extraordinary sharpness. Hovering before the base of the flower, the Fairy simply pushes its bill into the thick calyx and removes the nectar through the resulting perforation.

I was not easily convinced that, while so delicately poised in the air, the hummingbird could exert sufficient pressure to penetrate solid vegetable tissue somewhat over a twelfth of an inch (2 millimeters) thick; but after one had been visiting the flowers, I knocked down a number of the calyxes with a long stick and found many of them punctured by holes hardly wider than would be made by a common pin. I have watched hummingbirds of other kinds, including the White-eared and the Little Hermit, suck nectar from the flowers of various members of the mint family by forcing their bills through the bases of the long, tubular corollas while hovering beside them; but these hummingbirds perforated tissues much thinner and more delicate than the calyx of the poró. Flower-piercers—aberrant, often plainly attired highland representatives of the brilliant honeycreeper family of tropical America—regularly drink their nectar through perforations that they make in the bases of many kinds of flowers which are held steady by a hook at the end of the short, up-tilted upper mandible while the awllike lower mandible is forced into them. And bees, as is well known, sometimes bite little holes into corollas or nectariferous spurs whose length makes the sweet fluid difficult to reach with their tongues.

By piercing the bases of flowers, the Fairy extracts nectar from a number of species with corollas so large that it could not otherwise procure it, including red hibiscus flowers and white lilies with blossoms longer than itself. Unlike other short-billed hummingbirds, the Fairy appears to prefer big flowers; but whatever kind it visits, it perforates them instead of sticking its bill into the corolla's throat, where it might be dusted with pollen that it could transfer to other flowers. Thus, the Purple-crowned Fairy is a thief who fails to pay for its nectar by pollinating the flower, as most hummingbirds do, but it is a charming one.

Like other hummingbirds, the Fairy adds many minute insects and spiders to its diet of nectar. Possibly it extracts such solid no less than liquid nourishment from the Madera Negra flowers. One morning I watched a Fairy foraging at the edge of the forest beside the pasture in front of our house. Hovering before the foliage at the ends of exposed boughs, it seemed to be gathering from them objects too small for me to see. Then it darted erratically back and forth in the air above the pasture, apparently catching minute flying creatures—a habit widespread in the hummingbird family. Finally, it captured an insect large enough to be visible to me and carried it to a perch before eating it.

Late in January, years ago, I found my only nest of the Purple-crowned Fairy. It was about 30 feet (9 meters) above the ground in a small tree standing at the edge of a newly made clearing in the rain forest, close beside a rivulet and a woodland trail. The nest was saddled in the elbow of a nearly horizontal twig with a slight upward bend in the lowest tier of branches, far out from the center of the tree. In shape, the structure was a hollow sphere with the upper third, more or less, cut away to expose the central cavity. As far as I could see by examining it through my binocular, it was composed of downy materials, with none of the lichens or green mosses so frequent in hummingbirds' nests.

The female was incubating, probably the two minute, elongate white eggs usual in the family, although they were invisible to me. She sat with both ends of her slender body projecting well beyond the rim of her little chalice; on one side, her white throat was conspicuous to the observer on the ground; on the opposite side, the white under tail coverts and the lower surface of her long, tapering tail shone out. For a hummingbird, she sat very steadily. While I watched on a mild, bright morning, she incubated for sixty-eight minutes continuously, took a recess lasting eleven minutes, then returned to sit for fifty-six minutes. After her next outing, which lasted only six minutes, she came with a billful of down, which she tucked inside the nest, then ran her bill over the outer surface, smoothing it. This time she incubated for only three minutes, and when she returned after fourteen minutes, her bill was empty.

The Fairy's departures from the nest were spectacular, for on leaving she fell into the clear space beneath the crown of the supporting tree, tail spread and wings fluttering. Sometimes she dropped only a few feet, but once she fluttered downward almost to the ground. Finally recovering herself, she flew into the adjoining forest, never over the clearing. I have seen small birds of a number of kinds leave their nests by means of a sudden drop, which apparently would be less likely than a direct outward flight to reveal its position to an enemy which happened to have the nest in its field of vision. When she returned, however, the Fairy seemed to care little for secrecy, for she circled below the nest for a few seconds in a loose, fluttering fashion, her gleaming white underplumage most conspicuous, then deftly dropped into her cup's narrow opening. She was never accompanied by a mate.

From southern Mexico through the rainier parts of Central America to Colombia and western Ecuador, the Purple-crowned Fairy inhabits forest and adjoining clearings with flowering trees and shrubs. From warm lowlands it ranges upward to about 5,500 feet (1,675 meters) in southern Central America, where it has a long breeding season. The nest that I found in January was about a mile from our home in the Valley of El General. On Barro Colorado Island in Panama, occupied nests have been found in April and August (Sturgis 1928; Vleck 1981). In early September, also in El General, I watched two young Fairies who had not been long out of the nest; they were hatched from eggs laid probably in July. High in roadside trees, they rested on slender twigs and frequently flew around each other. Often they hovered with their short bills against leaves and new shoots, from which they seemed to pluck something that I could not identify. Presently their mother arrived and regurgitated to one of them while perching beside it. These juveniles resembled her but were smaller, with shorter tails. The whole top of their heads was brownish instead of green, and their backs were duller than in the adult. When she flew off, one of the young Fairies followed but the other stayed behind.

7. On Feeling Close to Nature

Emerging from the steep, forested slopes of the Cordillera de Talamanca where it is born, the Río Peñas Blancas flows for several miles between cleared farmlands before it reaches Los Cusingos. Despite its passage through cultivated country, it washes our shore as an untamed mountain torrent, its water as clear and sparkling, if not as safely potable, as when I first saw it many years ago. As it rushes down a boulder-strewn channel, it fills all the valley with its clamor. As I sit on the shore, watching intrepid Torrent Flycatchers flit from rock to rock barely beyond reach of the foaming surges or waiting for a solitary kingfisher to fly along its course, I find it wild enough to satisfy my contemplative mood. I am glad that it is not forbiddingly wild, as when, swollen by deluges in the mountains, its turbid water rises high, bearing trunks and branches, shifting the rocks in its bed with a deep rumble audible above its deafening roar, tearing into its banks, perhaps cutting a new channel through the farm. Then I stand somewhat apprehensively on a high bank, watching the brown torrent swirl by instead of sitting calmly on a streamside rock, grateful for its unspoiled beauty. It reminds me that, up to a certain point, wild, untamed nature soothes and uplifts the spirit, but excessive wildness, like all excesses, is disturbing and forbidding.

As with the river, so with the forest that begins within fifty yards of it. It is certainly not a vast wilderness but a modest tract of about a hundred acres surrounded by farmlands; but its trees rise as high as when I first knew them, palms tall and low flourish among them, epiphytes grow profusely on trunks and branches, the voices of birds punctuate its silence. In its midst, I feel the wonder of tropical rain forest hardly less than when I first came here. I do not regret the absence of Jaguars, Pumas, and dangerous bands of White-lipped Peccaries, all of which had disappeared before I arrived. I wish the Fer-de-lance and other venomous serpents had vanished, too. Whether I watch a bird's nest, study a plant, or try to interpret what I see of this profusion of life, the risk of attack by dangerous animals would distract me. Like the river, the forest offers most to the contemplative mind when it remains unspoiled but not too wild.

Standing calmly in the tropical forest, whose soaring trunks, dappled with stray sunbeams, lead my gaze upward, I feel as nowhere else the intensity of the creative energy that gives form to matter and fashions the immense variety of plants and animals around me—that fashioned me. I ask myself what is the meaning of all this grand display of organic forms, what the ultimate significance of the whole vast movement that produced it. It occurs to me that the experience which prompted the question suggests the answer. In just such moments of exalted awareness of the whole of which we are parts the cosmic striving appears to ful-

Torrent Flycatcher

fill itself. If I feel the wonder of life more keenly than the living things around me, it is not because I am infused with some principle that is lacking in them, but because the permutations of the evolutionary process have so organized me that I am more fully conscious of it. The thirst for a more abundant life that is perhaps dormant or only weakly felt in them has become stronger in me. Overflowing with gratitude for this priceless gift that I did not earn, I try to compensate for what I have and they appear to lack by grateful appreciation of all that they contribute to my exaltation.

Casting a more critical eye upon the life around me, I perceive that it is not all harmony and peace. Emaciated saplings and dying trees have failed in their struggle to find space in the forest's sun-bathed canopy. A fig tree has embraced the trunk that supports it with a maze of roots that will finally strangle the host. Perhaps a snake slithers through the undergrowth, relentlessly searching for eggs and nestling birds that it will swallow whole.

These things distress me but do not destroy my faith in the soundness of my intuition that the forest around me is a maximum expression of the cosmic striving to intensify consciousness and give ever higher value to existence. For I reflect that the silent strife around me is an inevitable consequence of the excessive intensity of a creative process that lacks foreseeing guidance. A moderating Intelligence might lead creation to fulfillment without all the waste, the misdirections, the bitter strife that afflict the living world. Here in the forest I feel, within myself and around me, so strong an urge to exalt existence that I cannot doubt its reality; but I find no indication of a compassionate guiding Intelligence. Life must find its own way upward, which is the reason why its progress has been so devious and slow.

I delight in the forms and colors of the plants around me, the fragrance of their flowers. I love the beauty of the birds and the varying tones of their voices. I show the wonders of tropical nature to visitors from northern lands who are fascinated and regret that their visits are so brief. Then I recall that recent explorations of the solar system by means of space probes support the conclusions of astronomers that the high degree of organization that surrounds me is rare in the Universe, perhaps confined to the planet on which I stand. In any case, only an infinitesimal proportion of the matter in the Universe can in any epoch attain such advanced organization. Somewhat more is at the crystalline level, but the vast preponderance is in the molecular or atomic state or is even less organized.

This situation leads some to conclude that we are cosmic accidents, apparently isolated in a vast, possibly infinite Universe where we will find no intelligent beings beyond Earth with whom we might communicate. This, too, fails to depress me while I stand amid such a convincing exhibition of the strength of the creative urge. Remembering that the same elements of which I am made are found throughout the Universe, and remembering the extreme rarity of environments such as this in which they can give full play to their capacity to create, I regard myself—and those like me—as the fortunate outcome of the cosmic striving to attain high levels of organization and awareness. I exult that in us the cosmos has achieved a fairly high level of success. Far from being cosmic accidents we are, in no small measure, the Universe triumphant in its aeonian striving to give significance to its existence, which includes our own. Instead of lamenting our isolation and the briefness of our mortal span, we should gratefully cherish every moment of our conscious life that we can fill with beauty and joy and kindness, for this is what gives value to the cosmos. If the life of this planet is unique in the Universe, its importance could not be greater.

My belief that the whole is of vast importance intensifies my efforts to learn the details. To the serenity that insight brings I wish to add the exhilaration of discovery. In the forest, as elsewhere, some valuable discoveries are made by chance; but the rain forest guards its secrets well, and their disclosure is usually the reward of sustained, concentrated effort. Not the most powerful of the sylvan animals, the great predators that dominate all the others, most repay our study. Many of the smallest and weakest timid creatures that move obscurely amid the verdure are the most finely organized, the most accomplished builders, the most faithful mates, the most careful parents. They are more adequate revelations of life's capacity for harmonious development than such truculent animals as raptorial birds, great carnivorous quadrupeds, or huge serpents that crush their victims and swallow them whole.

Evolution has been more successful in creating organisms that are marvels of biological efficiency than in developing harmonious relations between diverse creatures. Its highest, rarest, most admirable achievements are the associations of organisms which benefit one another without detriment to others. In the tropical forest, the association that includes flowering plants with edible fruits, frugivorous birds, and pollinating insects appears to be the most comprehensive and widespread. The plants offer nectar or nutritious pollen to bees and other animals. The flowers that they pollinate develop into fruits that attract birds, who digest the pulp and disseminate the seeds, many of which grow into new plants that perpetuate a cycle in which all participants benefit and none is injured. Such an association shows us what the whole living world might have become if evolution had taken a different course; it raises our estimate of life's potentialities.

Contemplating such an association, I speculate upon the possibility of making the whole living world one harmonious community of fruitfully cooperating animals and plants. If we had the power to create such a world, as by reforming certain predators and eliminating those incapable of changing their aggressive habits, would it not be our duty to do so? I believe that this course would be incumbent upon any moral being able to pacify the whole realm of life or any large segment of it, such as terrestrial life; but certainly a long age must pass before man could attain sufficient wisdom, knowledge, and power. Before we could undertake such a course, we must make a peaceful human world, of which the present prospects are not encouraging.

Although it would be ludicrous for man, in his present disordered state, to attempt to pacify the whole living community, it is hardly our duty to preserve all its most savage and destructive members. Evolution is an unguided process, dependent upon random mutations, in which survival depends upon nothing more than ability to live and reproduce, at whatever cost to surrounding creatures; certainly not every product of such a process is worthy of our protection. Bigness and predatory power do not give an animal a special claim to be preserved, as some conservationists appear to believe. We should be thankful that the great carnivorous dinosaurs vanished before man arose; we could hardly coexist with them. Similarly, one might name many another organism, from pathogenic bacteria and parasitic insects to venomous snakes and certain ravenous quadrupeds and birds, which might be permitted to go the way of the dinosaurs with no detriment to nature's beauty or balance, which could adjust to their absence as it has done to the extinction of many a member of the living community powerful in its day. Our resources for the vast undertaking of preserving Earth's flora and fauna are so inadequate that we should be more discriminating in their use, giving preference to those organisms which, by their peaceful coexistence with their neighbors, most deserve our protection. And the best way to protect them is by safeguarding their habitats.

To preserve most of the creatures that seem most worthy of preservation we do not need vast areas of wilderness. Modest tracts of forest or other habitats will suffice if they are large enough to support many of the original inhabitants, close enough to one another for some exchange of organisms between them to avoid excessive inbreeding, and can be protected from destructive human intrusions. A wilderness covering many square miles will retain a fuller complement of the original fauna than a small reservation, which seems inevitably to lose some species of birds and mammals despite our best efforts to save them. But a vast untrodden forest, far from the sources of necessities, rarely offers the most favorable conditions for studying nature. Most of our knowledge of organisms of tropical forests was gathered at the edge of the wilderness or in reservations of moderate size, such as Barro Colorado Island beside the Panama Canal in Gatún Lake or La Selva in northern Costa Rica.

Nature rewards us most richly when it is wild and unspoiled but not too wild. The rapprochement of man and nature requires adjustments on both sides. We must simplify our lives and seek closer communion with nature, while nature becomes more friendly and inviting by the mitigation of some of its perils. In some ways, I felt closer to nature in the rather depauperate, second-growth woods in which I roamed as a youth in Maryland than in the far richer tract of ancient tropical rain forest that for many years I have preserved close to my house. There, in the absence of dangerous snakes, I could walk through the nocturnal woods without a light, sit and sleep on the ground, as I would not care to do here, where venomous serpents, great spiders, and scorpions are most active at night. After barely avoiding the Fer-de-lance and other deadly vipers on several occasions, I rarely cease to be watchful for them. Such wariness appears to be habitual with most animals in their natural environments, but in orderly societies we tend to develop greater confidence in our safety—except on busy highways. Lurking perils make it more difficult to feel close to nature.

Nature can become friendlier in small reservations than in great areas of unaltered wilderness. Such tracts preserve certain values that cannot be retained in smaller areas; but they tend to be too remote and inaccessible, often too wild and perilous, for many people to enjoy them. What we most need is areas of unspoiled nature, of no great size but well preserved, close to our homes, so that we can enter them frequently to feel the wonder of life, be uplifted and refreshed. A landscape pleasantly diversified with dwellings and tilth in the valleys and forest on the hills, or with cultivated fields alternating with woodland, is no less beautiful than a great expanse of uninterrupted forest, where crowded trees narrow our view. Unfortunately, in these days of agribusiness and extensive monoculture, such landscapes become increasingly difficult to find.

Likewise, nature seems friendlier, less savage and alien, in the measure that we penetrate beneath its surface toward its hidden springs. The first step in this movement is to pass from the contemplation of animals in their maturity to animals in their infancy and childhood. Just as human children, innocent and trustful, often appeal to us more strongly than grown people who, in the hard struggle to advance themselves, may have become aggressive and suspicious, so young animals, harmless and playful, are frequently more winsome than adults of their kind. Their original nature has not yet been so heavily overlaid with the protective armor, the aggressive or defensive weapons, the fierceness and hostility with which evolution has burdened animals for their own preservation. They do not alienate our sympathy by savage attacks upon defenseless creatures. Because the character of young, growing animals has not been so greatly modified by the struggle to exist in a crowded world, we find it easier to be friendly with them.

As our thought penetrates from the surface where strife and struggle prevail to hidden depths where each individual life begins, we enter a region where we detect only harmonious growth, an advanced stage of the universal process of harmonization that set the planets in their courses around the Sun and unites atoms and molecules in glittering crystals. From the first cleavage of the fertilized egg to hatching, birth, or germination, and in many organisms well beyond this point, no suggestion of aggression or destructiveness mars the harmony of growth, which is fundamentally the same in all creatures, the plant and the animal, the least and the greatest, the mildest and the fiercest, the worm and ourselves. By similar methods it shapes the same stuff into the most varied forms with the most diverse habits. Here in the formative depths we may feel that perfect concord with nature which at the surface we seek in vain. The great tragedy of the living world is that organisms so similar in origin, formed by the same process, grow into creatures that too often relentlessly oppress and destroy one another.

Long ago, in a village that nestled in the shadow of a snowy summit untrodden by man, some youths, to prove their courage and stamina, challenged each other to a contest. Each was to climb the mountain alone and to bring back some token of the altitude that he had reached. The first to return brought an acorn; the second, a pine cone; the third, an alpine flower from the rocks above timberline. But the youth who was absent longest and climbed highest returned empty-handed to the village. He had nothing tangible to show, because the snow that he had gathered on the peak melted while he descended to warmer levels. But from the height that he had reached he had surveyed a wider expanse of mountains and valleys than any of the other villagers had seen, and he was ever afterward a more thoughtful man, sought by his neighbors for his wisdom. This parable, which I heard so long ago that I have forgotten its source, reminds us that the chief values which we derive from wild nature cannot be held in our hands.

8. Curious Flowers

About five years ago, a vine that I had never noticed before sprang up in many places at Los Cusingos, including the pasture, the banana plantation, and the garden. Only in the garden did it find supports on which to climb. Here its slender, dextrorsely twining, hairy stems wound up a young orange tree and spread their large, ovate-cordate leaves over the yellow and green foliage of a neighboring *Codiaeum* shrub. Soon it began to display flowers that helped me to identify it as *Aristolochia pilosa*, the Hairy Birthwort. These flowers are among the strangest that I have ever seen. They lack petals and consist largely of a most irregular calyx, all in one piece, about 2.5 inches (6 centimeters) long. From the swollen hollow base rises a long, tubular, ascending neck that flares out at the top, where it bears a long appendage of complex structure. Viewed from the side, the flower resembles a legless, long-necked bird with a great, thick beak. From the front, it reminds me of the head of a small mammal with prominent, rounded, deep brown ears and an exaggerated green snout bristling with coarse hairs. This flower yields no nectar and has at most a very faint odor, which a visitor described as woodsy.

This is one of the smaller of the flowers of the three hundred species of the large tropical genus *Aristolochia*. Over 1 foot (30 centimeters) in diameter, the flowers of *Aristolochia grandiflora* are so big that, according to Humboldt, Indian children in Colombia playfully pulled them over their heads as caps. The cordlike appendage at the apex of this flower may become over a yard long. The bloom of this and other species of birthworts diffuses a highly offensive odor of carrion. The roots of these vines are reputed to be lethal to pigs and other animals. I do not know whether our species is poisonous, but its leaves are extremely bitter, suggesting the presence of alkaloids.

The aristolochia in our garden blooms throughout the year, but most freely in the drier weather of the early months. The flowers open during the night. At dawn, the beak or tongue, which in the developing flower is rolled tightly inward, is spread out flat and hangs between the "ears." Soon it rolls up the other way, with the bristly inner surface outward—at first involute, it now becomes revolute. This transformation is completed around sunrise, when the flower is ready to receive its insect visitors. Each flower remains functional for less than two days. On the afternoon of the second day, the beak unrolls, flattens out, and hangs limply. Soon thereafter, the flower falls from the long, slender, inferior ovary.

The aristolochia flower is a trap to catch insects. The brown of the ears is not greatly different from that of *Stapelia*, a species of the orchid *Maxillaria*, and other flowers that are pollinated by carrion-feeding insects. Flying in through one of these furry ears, the insect finds itself in a tube lined with inwardly pointing hairs which facilitate its entry but prevent its exit. Passing through a ring of stiffer inwardly directed hairs at the lower end of the tube, the visitor emerges into an egg-shaped chamber, with colorless hairs thinly spread over the translucent walls. On the chamber's ceiling, near the forward end, are two dark patches, one on each side of the midline. Each patch is composed of dense tufts of long, delicate, thin-walled hairs that resemble certain filamentous algae. From the base of the chamber rises a low column bearing at its apex six U-shaped stigmas arranged in a circle, with tiny hooked hairs on their receptive surfaces. Six two-celled anthers surround the base of the column beneath the stigmas, with which they alternate.

The pollinators of these peculiar flowers are tiny black flies, about a twelfth of an inch (2 millimeters) long. Some enter the newly opened flower with little mounds of yellow pollen conspicuous on the hairy top of the thorax. The sticky stigmas are waiting to receive these grains, but the anthers are still tightly closed. The little pollinators remain captive until the second morning, when the anthers shed their pollen and the inwardly pointing hairs that barred their exit wilt, permitting them to escape, many bearing loads of fresh pollen that they may convey to other flowers. They leave minute, elongated white eggs on the stigmas. In return for the flies' services in effecting cross-pollination, the aristolochia offers them a place for the development of their larvae.

This exchange of benefits seems fair enough, but, as frequently happens in the living world, a mutually beneficial arrangement between two species is complicated by the presence of a third, a shiny black rove beetle up to three-eighths of an inch (1 centimeter) long. The insect's slender, flexible abdomen extends well beyond the short elytra and can be bent forward to groom its back, head, and wings with the pointed tip. These staphylinid beetles are at least as abundant in the flowers as the flies; I have found up to twenty in one chamber. No pollen sticks to their glossy bodies, and they do not lay eggs in the flowers. They appear to enter chiefly to eat the trapped flies, whose mangled remains and detached wings often lie in the chamber where the rove beetles are present. Since I did not see a beetle catch a fly, I cannot be sure that those which they devoured had not died from other causes; but the abundance of their remains in certain flowers and the presence of injured living flies, together with the known predatory habits of certain staphylinids, lead me to believe that they prey upon living victims. I also attribute to them the mutilated state of the stigmas and anthers of some flowers.

In an effort to learn the function of the twin patches of brown filaments on the roof of the chamber, I cut them out with a small piece of the surrounding tissue and placed them in the glass tube of a pipette with the large end open. I wrapped a green leaf around the tube and placed it on the leaf of a flowering aristolochia vine. I suspected that these filaments might produce a scent attractive to the flies and the rove beetles which would draw them into the tube. The results were negative; I never found an insect in the tube.

Noticing that the tufts of filaments are sometimes partly eaten, one morning I tried a different experiment. Cutting open a flower to expose the tufts, I placed it in a transparent box with about a dozen flies and two of the black beetles. At first, all the insects tried to escape toward the light, but after a while the flies came, singly or two together, to the tufts, apparently to eat the filaments or something found on them. They are so small that I could not distinguish details, but their tiny bodies moved vigorously while they dined. After several hours in the box, the flies, apparently becoming hungrier, alighted on the tufts in larger numbers, and presently the two beetles joined them. For a long while, the flies and beetles ate peaceably side by side, until by early afternoon much of the tufts had been devoured. Apparently, the filaments serve to nourish the insects during their long incarceration. When tiny ants found their way into a box with a flower that I had cut open, they clustered around the tufts of filaments until they had consumed nearly all of them. They also ate hairs in the tube, but not so consistently. I lacked reagents to determine the nature of the deep brown granules, embedded in a matrix of lighter brown, that the filaments contain.

I have found flies and beetles in the flowers in every month and eggs or larvae of the former in all months except

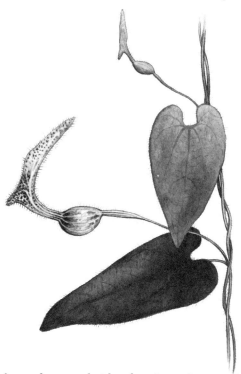

Aristolochia pilosa *columns (top left) viewed from above and side, showing stigmas with fly eggs and anthers; rove beetles (bottom left) prey upon the flies. At right, leaves and unopened flowers.*

May, August, and September. The eggs were all of one kind, roughly ellipsoidal, white-bodied, about .6 millimeter long, densely covered on one side with short projections that expand upward. They hatch in a day or two and the minute, colorless larvae crawl over the stigmas, apparently deriving nourishment from them. Before they can pupate, the flower falls and decays, leaving the larvae to complete their growth and pupate in the ground. Since they are difficult to follow there, I placed earth in the bottom of a glass jar, added flowers in which eggs had just been laid, and covered the top with a cloth. After two weeks flies began to emerge, and more followed on subsequent days. The fly's development from egg to mature insect accordingly requires no less than fourteen days. No beetles ever emerged in the jar, and flowers that held many beetles but no flies never contained eggs—all of which strengthened my conviction that they never lay in the flowers.

The birthwort flowers are followed by ribbed green pods about 2.5 inches (6 centimeters) long by .75 inch (2 centimeters) broad. In dry weather, the end of the pedicel splits into six stiff strands which bend outward and upward, forcing apart the bases of the pod's six valves while their apices remain attached together, thereby forming a sort of open

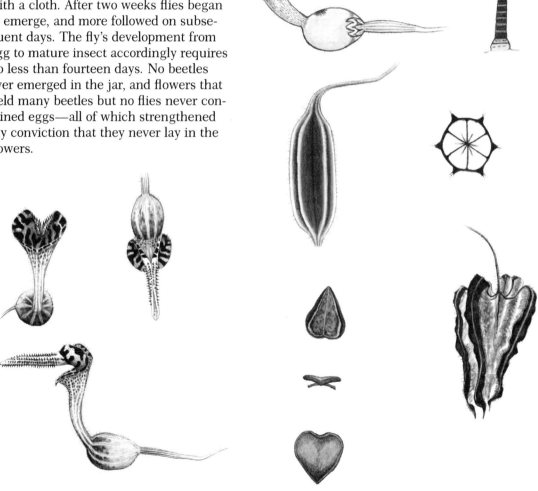

Aristolochia pilosa flowers (left column) viewed from front, above, and side. Right column, top to bottom: flower cut open lengthwise, with a stiff hair, enlarged, from entrance tube; pods, whole and in cross section; seeds and open pod.

basket shaped like a top or inverted cone. From seventeen to twenty-seven of the very thin seeds are stacked in a single long row against the inner face of each valve, held there by thin, incurved, membranous edges derived from the partitions between the pod's six cells. Here they remain until they are shaken or fall out, which in dry weather occurs rather promptly. In wet weather the pods do not open to form baskets, but the valves simply split apart and hang limply side by side, making the dispersal of the seeds less effective.

The thin, heart-shaped seeds are about one-sixth of an inch (4 millimeters) wide. Unlike certain other species of *Aristolochia*, these seeds lack broad wings; but on their back are curious thin attachments that may be rudimentary wings. Probably their very thinness and lightness permit their transport by the breezes, for new plants spring up in open places at a distance from flowering vines. The earliest of the seeds that I planted in sand took a month to germinate, while others delayed three and a half months before sprouts appeared.

Another plant that bears unusual flowers is *Mucuna holtoni*, one of the vines called in Costa Rica Ojo de Buey (bull's eye) because of the size and shape of its seeds. I have found it chiefly twining up trees at the forest's edge or in light second-growth woods up to at least 3,000 feet (915 meters) above sea level in El General. A member of the Leguminosae or bean family, it blooms during the rainy months, from May until November, starting to flower earlier and finishing sooner in some years than in others. From the axil of one of the compound leaves with three ovate leaflets springs a cordlike peduncle, which continues to grow downward until it attains a length of from 4 to 9 feet (1.2 to 2.7 meters) and bears its flowers dangling freely in a clear space, a most unusual arrangement even in the tropics, although some Old World species of *Mucuna* display more brilliant flowers at the ends of much longer strings. The compact inflorescence contains about a hundred flowers, springing in groups of three from the thickened lower end of the cord. Pale green throughout, about 1.75 inches (4.5 centimeters) long, these flowers have the typical papilionaceous structure of standard, keel, and wings but their behavior is quite different from that of most leguminous flowers.

The flowers open downward from the base, or top, of the inverted inflorescence. A full-grown bud hangs with downwardly directed apex at the end of its more or less horizontal pedicel. It opens in two stages, the first of which is the lifting of the standard. In deep shade this may begin before daylight fades, but it occurs chiefly between seven and eight o'clock in the evening, when the last traces of daylight have faded from the sky, and sometimes even later. The standard's outward and upward movement is at first imperceptibly slow, but after it is half raised, it completes the process by a sudden and very obvious upward sweep. In from three to seven minutes after it begins to separate from the rest of the flower, the standard is fully raised and stands out horizontally, like a green hood, from the front of the elongated flower, while the stamens and pistil remain within the tightly folded keel.

The flower is now "set," and it remains in this condition until sprung by an appropriate contact, such as the insertion of the point of a pencil between the appressed edges of the keel where it emerges from the calyx. When this occurs, the keel springs open and backward, the stamens spring forward, releasing a little cloud of pollen, and simultaneously the style bends forward and upward. The flower can also be sprung by gently pressing the sides of the calyx between one's

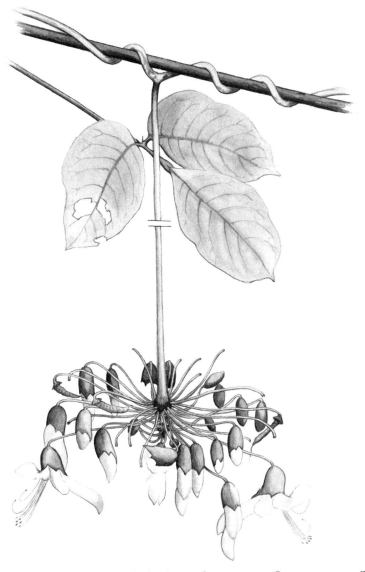

Mucuna holtoni *above and below, from left: bud, pistil, unsprung flower, sprung flower*

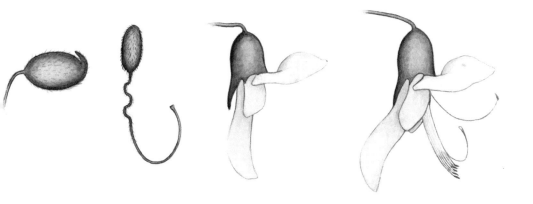

fingers. The newly opened flower generally has a glistening drop of nectar in its throat.

These scentless flowers, devoid of color that contrasts with that of foliage, hanging freely in midair, appear to invite pollination by bats. Other species of *Mucuna* are known to be bat-pollinated (Baker 1970). Watching these flowers at various times during the night, either by moonlight or with intermittent illumination by a flashlight, I have seen only a variety of insects visit them, all too small to spring the keel or effect pollination. Many flowers remain set until morning, when they are visited by nectar-drinking birds. Those that I have most frequently seen are Bananaquits. Clinging above a flower, the tiny, yellow-breasted bird pushes his sharp, curved black bill into the keel, making it spring open and shoot toward him a cloud of whitish pollen, conspicuous in sunshine. He comes again and again, on each visit probing several flowers, until nearly all are sprung. Between visits to the flowers, he sings. But he is poorly situated for receiving pollen on his body or transferring it from flower to flower; only exceptionally does his neck appear to brush against a stigma. Among the few species of hummingbirds that I have seen at the *Mucuna* flowers is the Purple-crowned Fairy, which commonly pierces the bases of all kinds of flowers and pollinates few, if any. I have not seen the Fairy release the stamens. If the inflorescence hangs low, it may be visited by Little Hermit Hummingbirds, who sip nectar without piercing the flowers but always so swiftly that I have been unable to distinguish details.

Newly opened flowers that I covered with gauze to prevent access to them by birds or other creatures remained unsprung until they fell about thirty-six hours later. If they remain long unvisited, they do not spring open, releasing the stamens and pistil as a result of increasing inner tension, and apparently they are not self-pollinated. The small proportion of flowers that set seed, even if they are tripped by animals that do not transfer pollen from other flowers, also points to the conclusion that these flowers are not self-pollinated. From the hundred, more or less, flowers in a single inflorescence, rarely more than three or four seed pods are formed; the largest number that I have found is seven. Since so few flowers appeared to have been visited by effective pollinators at the plants that I watched, it is not surprising that I failed to see a bat come to them.

The length of the pods depends upon the number of seeds they contain. The largest that I could reach had five seeds and was 9 inches long by 1.5 to 2 inches broad (23 by 4 to 5 centimeters). Most have fewer seeds and are shorter. The pods are swollen around the seeds and constricted between them—torose, as botanists say—like those of Beggar's Ticks and the Red Poró. As they dry and turn brown, the pods split open, releasing from one to six seeds which are black and nearly smooth but not glossy like the *Mucuna* seeds ("sea beans") cast up on tropical shores. When thoroughly dry, they are an inch or a little more in diameter by nearly half an inch thick (25 by 12 millimeters). Unlike the pods of some species of *Mucuna*, those of *holtoni* are devoid of stinging or irritant hairs. The outside is velvety with a very fine, dense pubescence and the inside is hairless.

Oddities in the floral world are the flowers of certain species of *Clusia*, which, instead of attracting their pollinators by nectar and pollen, offer them resin. *Clusia* is a large genus of shrubs and small trees confined to tropical America. Most of its species are epiphytes that start life on trees and rocks. Sometimes, after the decay of the host tree, they manage to hold themselves erect by thickening the roots that have grown down the supporting

Clusia *inflorescence with open staminate flower and (below) staminate flower in longitudinal section, showing tight cluster of anthers*

trunk into the ground; rarely they appear to start as self-supporting seedlings. When cut, their trunks exude a whitish or yellowish latex. Their leaves are usually thick and glossy. Like many tropical trees, they do not form resting leaf buds enclosed in scales, but the bases of the petioles of the terminal pair of opposite leaves, bending inward until they press against each other, tightly enclose the rudiments of the next pair of leaves. The flowers of clusias are most varied.

The most abundant species around our house is *Clusia uvitana*, which is best described as an overgrown shrub with very long, stout, ascending branches; rarely it is a tree with a single trunk. Some bear only female flowers, some only male, but with age the latter may produce pistillate flowers and fruits. The flowers of both sexes, about .75 inch (20 millimeters) broad, have five rounded, fleshy petals, glossy red with broad whitish margins. They secrete no nectar and are devoid of fragrance. In the center of each staminate flower is a fleshy mound of tissue with about twenty-four minute anthers situated around its margin. Its rounded top is thickly covered with an extremely sticky resin. In the center of the female flower rises a higher mound, the ovary. Its top is occupied by six to eight sessile stigmas, beneath which the mound is encircled by a glandular ring that secretes a resin similar to that of the staminate flowers.

Of a morning, I have watched many small, stingless bees of several species alight on the resin, collect it in tiny pellets which they stick on the pollen baskets on their hind legs, then fly away with full loads. Surprisingly, they do not stick to this extremely adhesive stuff, which is most difficult to remove from a knife or one's fingers. The bees use the resin to make the roughly globular vessels in which they store honey and pollen and the spouts that project from the sides of hollow trunks in which they live. They also employ it most effectively against invading ants, whose movements it greatly impedes, as I described in *A Naturalist in Costa Rica* (1971). Bees appear to be the only insects that pollinate this clusia, and they do so efficiently, as it sets many fruits. When mature, these open by six or seven valves that spread like the petals of a flower, exposing many tiny seeds embedded in bright orange arils that birds of many kinds eagerly seek, as described in *Nature through Tropical Windows* (1983). After digesting the soft arils, the birds void the seeds, often on trees where they start new clusia shrubs.

A different clusia began life on the trunk of a tree fern in a sunny spot in front of our house but finally became well rooted in the ground. It has large, thick, deep green oblong or obovate leaves and flat, white, staminate flowers, about 1.5 inches (4 centimeters) wide. These flowers open during the night, and by early dawn the five broad, rounded petals have spread far enough to reveal in their midst a low mound of tightly packed stamens. The broad, slightly convex top of this mound is covered with amber-colored resin, about the color and consistency of extremely thick and sticky honey or treacle. Above each anther cell this resin stands up as a minute dome; the whole top of the mound is densely studded with miniature cupules.

Before five o'clock, while the dawn light is still dim, stingless meliponine bees arrive to gather this resin. These bees are chiefly the little black Aragres (*Trigona* sp.) that build on trees and houses large, blackish, roundish nests that resemble arboreal termitaries but can be distinguished by the funnel in the side through which a stream of bees flies in and out. I have counted as many as eight of these bees working together on a single flower, accumulating great masses of the resin on their hind legs. The round lid that closes the top of each anther cell appears to be pulled off as the resin covering it is removed, releasing abundant pollen in the form of smooth, spherical grains. The bees appear to be attracted by a very faint, honeylike fragrance, which lured them to flowers on my table inside the house. Bees appear to be the only visitors to these flowers devoid of nectar. The pollen is apparently carried in the resin; but it does not send forth pollen tubes while it remains in or on the resin, whether on the mound of anthers or on a glass microscope slide. By seven o'clock on a bright morning, the bees have emptied most of the anther cells, leaving the top of the mound covered with tiny white depressions, each surrounded by a rim of honey-colored resin. Each flower lasts a single day. By the following morning it has turned brown, and soon it falls. I have not found a female plant of this clusia, which probably grows more abundantly high on the great trees of the forest.

The foregoing are all native plants that spring up spontaneously at Los Cusingos. The last of the curious flowers of which I wish to tell is a Brazilian species planted in our garden, the gift of a neighbor. *Marica caerulea* is a vigorous member of the iris family whose long, narrow, erect leaves stand 4.5 feet (1.4 meters) high, while the tips of the broadly winged flowering stems rise to over 6 feet (2 meters). The lovely flowers, 4 inches (10 centimeters) wide, have three broad, flat violet or pale purple sepals and three curiously curled petals that are deep purple with white central streaks. The bases of all six of these perianth divisions form an open cup narrowly banded with light brown and pale yellow and thickly covered with tiny hairs tipped with glistening, colorless droplets that may be the sources of the flowers' delicious fragrance. In the center of this cup three stigmas and three stamens are joined in a compact, erect column.

These flowers open in a peculiar manner. Early in the morning, all that will expand that day stand conspicuously as pale lavender conical buds, 2 inches (5 centimeters) high, with sharp yellowish tips. Remaining attached at the top, the sepals bulge out below, forming gaps in the bud through which the purple petals are revealed. Suddenly one of the sepals breaks loose and bends rapidly outward. Still connected at their tips, the other two sepals remain together usually for several minutes longer, then break asunder and spread out in a trice. One must watch closely not to miss these rapid movements, which occur chiefly between seven and half past eight o'clock in the morning. Soon the flowers diffuse a sweet and delicate aroma, but they appear devoid of nectar. They attract little black *Trigona* bees with amber-colored abdomens who often appear confused by these alien blossoms and, after hovering around or alighting upon them, leave as empty as they came. Others collect pollen from the long narrow anthers. The flowers wilt soon after the middle of a sunny afternoon; in cloudy weather they may remain fresh somewhat longer.

Although most herbaceous plants open some flowers daily throughout their season, *Marica caerulea* blooms only on special days. After a generous display of thirty or more large flowers on a certain day, our patch of *Marica* remains wholly flowerless for two to four (rarely five) days, until on the third to fifth morning it opens another set of blossoms. Rarely, smaller numbers of flowers open on two consecutive days, as though some of the buds had fallen out of step. Thus, last year, throughout August and September, the main period of bloom, the plants flowered on only eighteen days. In October, when the period of anthesis was coming to an end, once nine days of heavy downpours elapsed between displays of flowers and twice the flowerless interval was six days.

The only other herbaceous plant with similarly synchronized blooming known to me is the white-flowered orchid, a species of *Sobralia*, of which I wrote in *A Naturalist in Costa Rica*. Its flowerless intervals were longer, sometimes two or three weeks. Among shrubs and trees that exhibit synchronized or gregarious flowering are coffee and numerous species of *Miconia* in the melastome family.

9. *The Appreciation of Nature as a Unifying Power*

Living is essentially a lonely affair. Life cannot be shared or transferred, in the sense of giving part or all of our vitality to another whose strength is failing, as, in one of the most beautiful of the Greek myths, Alcestis did to her husband, King Admetus. Each person's life depends wholly upon the beating of his own heart and the adequate functioning of his own vital organs—a limitation only precariously overcome by the acme of surgical skill, which replaces weakening organs by sound ones from another body, although the recipient body tends to reject them. At least for the more advanced animals, communal living extends only to externals. The indispensable vital functions that keep us alive are strictly individual.

Each of us is an insulated being, not only physiologically but likewise psychically. Our psychic life is hidden from those around us, as theirs is from us. By speech, facial expressions, gestures, and the like we try to convey our thoughts and feelings to our companions; but the mental states so induced in them rarely have the clarity or intensity of the originals, and misinterpretation is frequent. Our efforts to interpret the minds of animals, even those most closely associated with us, are groping and hazardous. About the feelings of creatures more distantly related to us, of invertebrate animals, plants, and perhaps even lifeless things, we can only vaguely conjecture. Astronomical distances often seem to separate mind from mind. This insulation, this difficulty of making oneself understood, is often distressing to sensitive spirits. To the philosopher, who strives to interpret and evaluate the cosmos, the obscurity that envelops its psychic aspect, which alone can give value to existence, is a source of perpetual frustration.

Our insulation is no accident but a basic condition of life. A living organism is inevitably a strictly delimited fragment of the Universe. To preserve its intricate physiological processes, it must exercise a high degree of control over what enters its body and what escapes from it. Even the simplest organisms, active only in water, such as the amoeba, the paramecium, or the euglena are separated from the medium by a semipermeable membrane which regulates the inflow of nutrients and prevents the outward diffusion of essential substances. Loss of this semipermeability is fatal.

When, from its cradle in the sea, life crept forth upon the land, it needed more adequate insulation from the environment, for death by desiccation was a recurrent threat. Every advance in organization demanded better insulation; trees exposed to drying sunshine and wind need better safeguards against water loss than mosses that grow in sheltered moist dells. The most highly organized members of the living community, mammals and birds, have achieved a high degree of control over their heat exchanges with their surroundings, so that they can maintain a nearly constant body temperature and remain active, despite great fluctuations in the ambient temperature. More efficient insulation has been one of the most important features of evolutionary advance.

Insulation from the environment includes insulation from other organisms, which are influential components of each individual's environment. Living things appear to have been condemned from the beginning to isolation or separateness, from other individuals and from the Whole, a situation which many, especially mystics, have regarded as a source of evil, perhaps its main cause. To preserve our insulated existence, we must resist so much yet take so much, for environmental extremes of any kind might destroy us; and to feed, clothe, and house ourselves we destroy other organisms, plants if not animals. Complex animals cannot mingle intimately with almost everything else, like some simple substance such as water, which, as Lao-tzu said of old, benefits everything and harms nothing. Most lamentable of all, our insulation, physical and psychic, from organisms of all kinds is an obstacle to the understanding and sympathy which would help us live in concord with them.

Although not without its darker side, the insulation that is a fundamental condition of our existence as living organisms is precious to us, for it is the foundation of our individuality. Only as individuals, localized in space and time, can we survey the world with a definite perspective, think, love, aspire, and try to transcend, on a higher plane, the separateness to which we are irrevocably committed at the physiological level. The mystic may strive, perhaps successfully, to overcome his separateness by turning his mental vision inward, but this path is too steep and lonely for most of us to follow. For us, the best remedy for oppressive separateness appears to be appreciative awareness of other beings. Let me explain by telling of an experience which, like all our most memorable experiences, was unique yet of a kind that everyone responsive to nature's grandeur, beauty, and infinite diversity may sometimes have.

It is a brilliant morning in the dry season, and I am reclining, half-hidden, amid flowering herbs high on a Guatemalan mountainside. In front of me, across a seldom-used roadway, is an odd, coarsely branched shrub of the potato family, adorned with large pale lavender flowers. Beneath a cluster of these pretty blossoms hangs one of the most exquisite birds' nests that I have ever seen, a pear-shaped pouch, about 6 inches (15 centimeters) long, composed of finely branched gray lichens, neatly arranged and bound together with cobweb and fine projections from the edges of the lichens themselves. The entrance to this dainty nest is a round doorway at the top, just below the flowers. A pair of tiny gray Bushtits, relatives of the chickadees and titmice, are busily engaged in lining the pouch with down, gathered chiefly from the large, furry leaves of the lavender-flowered shrub. The black-faced male and his brown-cheeked mate are working in closest harmony, usually coming together with little tufts of down or cobweb to bind them more firmly together. As each in turn enters to attach its contribution to the wall, the pouch shakes and sways so violently that I fear it will be torn asunder; but the Bushtits will not destroy the fabric on which they have already worked so long and arduously. The task of incorporating the fresh material into the wall completed to their satisfaction, the tireless birdlings fly off together to search for more.

The background of this endearing drama of animate life is on the grandest scale. Far below me lies the high plateau of Chimaltenango, where white-walled towns stand amid brown fields that await the return of the rains to be planted with wheat, maize, and potatoes. On and on, in misty vastness, the tableland stretches to great volcanic cones, rising clear and sharp into a blue sky. A wisp of smoke ascends from the crest of Fuego. In a single view, I have nearly all that one can ask of nature: the loveliness of its flowers, the charming spectacle of gentle creatures working closely together to create a beautiful structure, the vastness of its landscapes, the grandeur of its mountains, the works of man, too distant to produce a jarring note—all enjoyed amid bracing mountain air.

And what long ages, what stupendous forces, what intricate adjustments were needed to prepare this spectacle for my enjoyment, to prepare me to perceive and delight in it! What heavings and convulsions of Earth, what cataclysmic eruptions of volcanoes like those that rise before me, what slow erosion by wind and rain were needed to lift and sculpture this mountainous land. What an inconceivably complex evolution was required to produce the varied vegetation that covers this land and the animals that inhabit it, including the flowering shrub and the birds building amid its blossoms. And all this vast preparation might never have attained its culminating significance if another evolutionary trend had not equipped me—and others like me—with senses keen to perceive all this beauty and grandeur, with minds to appreciate it.

In an experience like that which I enjoyed many years ago, many unrelated developments, such as the titanic forces that lift mountain ranges, the evolution of plants and animals, the physical and intellectual development of man, converge to a focus of significance. Things as diverse as a tiny bird and a gigantic volcano, a flowering shrub and the incandescent star that illuminated the wide landscape and supplied energy for all the life around me, including my own—all contributed to a single unifying experience that emphasized the glory of our planet and the importance of its living cargo. Possibly the Bushtits enjoyed building their cozy nest and the shrub felt gleams of satisfaction as it absorbed sunshine in its downy foliage, but I doubt that either noticed the distant volcanoes, and certainly neither thought of the long ages of slow evolution to which it owed its advanced organization. Only an appreciative human mind could bind all this diversity together in an act of grateful acknowledgment that lifts existence to a higher plane.

I doubt that bird or plant benefits directly by being observed and appreciated, although indirectly each might do so if appreciation leads to protection. Certainly a mountain peak gains nothing by being admired from afar. Nevertheless, the value of the whole is augmented by appreciative contemplation. Bird and plant and mountain and man, interacting to increase appreciation, form a manifold whose value exceeds the sum of the values of these same four entities insulated one from another or related only by physical forces. Although appreciative contemplation cannot annul organic insulation, it overleaps its boundaries and unites us ideally with all that excites our love or admiration and makes us feel less alone. A world so bound together by an appreciative mind is certainly superior to one in which everything exists only for itself. And the more such minds a planet bears, the more precious it becomes.

Scarcely anything so binds mind to mind as shared admiration or appreciation. If, while watching the birdlings build their nest on the flowery mountainside, I had been accompanied by another who appreciated the experience as profoundly as I did, the number of entities unified by it would have been increased by one, but the value of the whole would have been more than doubled, augmented by the sympathy between minds.

The fact that people capable of deep appreciation do not all appreciate the same things is, in one aspect, fortunate, for no finite mind can appreciate more than a small fraction of Earth's wealth of things worthy of love and admiration. Diversity of interests and tastes not only enriches society but ensures that fewer things able to excite appreciation are neglected. From another point of view, the fact that we do not all appreciate the same things is lamentable; if we did, people of all races and lands would understand each other better and perhaps live in greater concord. Appreciation of appreciation might bring us closer together. The person who delights in orchids appears to have little in common with one who, for example, is enchanted by minerals and seldom looks at orchids. Nevertheless, the fact that both are capable of high enthusiasms might be a bond between two understanding minds. This, however, appears too sophisticated an approach to the problem of intensifying people's feeling of brotherhood by means of a common object of appreciation. Is there nothing that everybody should and could appreciate?

Obviously, to be appreciated by everybody, the object must be accessible to everybody and contribute to everybody's enjoyment or welfare. Such an object is Earth that supports us all. Widespread, adequate appreciation of our planet, now so undervalued and abused, would not only contribute greatly to its health but bind people of all races and nations together in a shared purpose and ideal as nothing else could.

Only by the united efforts of people everywhere is there a fair prospect of saving from ultimate ruin a natural world that knows nothing of the boundaries that men have artificially drawn between nations or the jurisdictions that they have more or less arbitrarily established. These boundaries hardly affect the planetary circulation of air and water. Unless they coincide with natural barriers like wide rivers or high mountain ranges, they pose no obstacles to the dispersion of plants and animals. Yearly it becomes more apparent that even the attempt to control population, humanity's most urgent need, cannot be effective within the boundaries of a national state. Despite immigration laws, the country that achieves a low birthrate is invaded by the overflow from more densely populated, less prosperous countries and the beneficial results of its own people's restraint in reproduction are canceled by irresponsible aliens. Clearly, only the united efforts of people everywhere can save the most favored planet in the solar system from being overburdened and overexploited by teeming humanity. But before people will care for it as it deserves, they must appreciate it as it merits.

Nobody with adequate appreciation of his planet's uniqueness, beauty, and bounty would wage war, for such appreciation would override all greed for territory, power, and wealth. Ancient warfare killed people, destroyed cities, and reduced whole populations to slavery, but its impact upon the natural world was apparently slight and local. Modern technological warfare wreaks inestimable havoc upon nature on the largest scale. No matter which of the belligerents wins, the planet loses. War's incredible waste drains all the world's resources: its forests are felled at an accelerated rate to supply timbers for military purposes; its minerals are mined in vast quantities for armaments and ships that are soon destroyed; its oceans are polluted by oil from sunken vessels; its face is scarred by high explosives and poisoned by chemicals, including radioactive substances; remote islands are invaded for military bases or testing grounds for atomic bombs and their unique organisms decimated or exterminated. All nature suffers when nations, forgetting what they owe to Earth that supports them, attack each other with barbaric fury.

The conventionally devout try to foster the feeling of brotherhood by reminding us that all people are children of one God. Unfortunately, monotheistic religions, with their diverse concepts of Deity and different schemes for salvation, have been responsible for the most irreconcilable enmities and the most cruel persecutions. The motherhood of the visible, tangible Earth, so generous to us all, might become a stronger bond between all peoples than the fatherhood of an invisible God, difficult even to conceive of convincingly. The determination to honor and protect our bountiful mother might unite us all in a common endeavor. To know that people in the north, thousands of miles away, feed and protect the same migratory birds that I feed and protect in their winter home in the tropics makes me feel more kindly toward them, although I am ignorant of their names. To know that in many countries a dedicated few are struggling to preserve the natural world against increasing pressure from hungry or greedy multitudes makes me feel less alone. If appreciation of Earth became general, instead of pitying humanity and despairing of its future we might all be drawn together in a hopeful fellowship of caring for the source of all our blessings.

To adequately appreciate the wonder of our planet, we must view it in its cosmic setting, as we too seldom do. Our ancestors, who believed Earth to be the only world created by an omnipotent God in a moment of his eternal existence, lacked the background for viewing our planet in true perspective, for which we are indebted to the discoveries of modern science. Their admiration and reverence were centered on the Deity who could create the world with so little effort, as though to amuse himself; the wonder was that why, being so wise and powerful, he did not make it more perfect, with less violence and pain. But when we view this terrestrial sphere as hardly more than a speck of dust on the astronomical scale, set amid such a vast extent of space, amid so many billions of huge, flaming suns, and we behold it covered with an immense variety of beautiful living forms that no incandescent star could support, including some that feel and think and aspire, our wonder increases a thousandfold. That such a minute fraction of the Universe should support so much of all that, as far as we can tell, gives value and significance to the whole of creation arouses endless wonder.

Doubtless, many of the billions of stars in our galaxy and others have life-bearing satellites, some of which may nourish beings far more intelligent or spiritual than ourselves. But all the inhabited planets together can comprise no more than a tiny fraction of universal matter and occupy an almost negligible proportion of space. Just as the importance of our eyes is not to be measured by their minute fraction of our total weight, so the importance of life-bearing planets cannot be gauged by their paltry mass and volume. Planets inhabited by intelligent, appreciative beings might be regarded as the eyes of the Universe through which it sees itself, the minds of the cosmos by which it strives to know and understand itself, the conscience of creation through which it passes moral judgments on itself. Life-bearing planets are the expression points of the world process which appear the more wonderful the more we contemplate them as minuscule bodies hardly visible amid the cosmic vastness.

It is our privilege and duty to ensure that one of these planets, the only one of whose existence we are certain, continues for long ages to remain a fit abode for intelligent, appreciative beings. To make this the common purpose of people everywhere would unite mankind as never before.

Appreciation is our best resource for transcending biological insulation. It draws us closer to whatever delights us or wins our admiration. Things the most diverse—the Sun and a flower, a bird and a mountain—that together contribute to a single experience which lifts the spirit above its humdrum daily level are bound together in an ideal unity of significance. We are strongly drawn toward those who unselfishly appreciate whatever most delights us, so that shared appreciation becomes one of the strongest bonds of fellowship. The best hope for the solidarity of mankind is widespread, cherishing appreciation of the planet which, properly cared for, could support and embellish the lives of all its human inhabitants. The achievement of such unity of purpose would immeasurably increase the happiness of people everywhere.

10. *A Village in the Treetop*

One who has watched, even casually, a pair of birds build their nest, incubate their eggs, and raise their young can best appreciate the absorbing interest of watching the busy life of a colony of oropendolas, centering around eighty pendent nests in the same lofty treetop.

When I returned to the Lancetilla Valley in northern Honduras late in April, the Montezuma Oropendolas were already well advanced in their nesting in the same tree where I had found them breeding the preceding year, for they flock each spring to the favored tree, as sea birds congregate from afar to lay their eggs and rear their young on the same barren islet where they have nested for many years. The tall, smooth-barked tree towered above an almost impenetrable thicket of low bushes and tangled vines that had overgrown an abandoned banana grove. On one side the narrow strip of thicket was bordered by a well-worn path that led to the thatched huts at the head of the valley; on the other, beyond a little grass-choked rivulet, a hillside pasture rose steeply to the west. Its slope commanded an excellent view of the nest tree, whose upper boughs were already laden with threescore hanging nests, clustered near the ends of the twigs like great, gourd-shaped fruits. One unfamiliar with the ways of tropical birds might have been surprised to see a big, chestnut-colored bird with a black head and neck and bright yellow outer tail feathers suddenly emerge from near the stalk of a seeming fruit, to fly with measured wing strokes toward the abrupt, forest-clad mountains that enclosed the narrow valley. In these nests the incubation of the two white eggs was still in progress, or the young had already burst their shells and were crying loudly for food.

But a group of twenty-one oropendolas, more sociable than prudent, had clustered their twenty-one long nests among the twigs of one slender living bough. Everything had apparently gone well with them until many had laid and started to incubate their eggs, when the overladen

Montezuma Oropendola male displaying at nests

branch had snapped off at a point where it was two inches thick and crashed down to the brink of the rivulet. All this had happened a week or more before I arrived to find the foliage already withered on the fallen branch and the nests discolored by dampness. Since examination of the contents of the nests furnished no evidence that either parents or young had lost their lives in the accident, I was not altogether sorry about it, for it provided an opportunity to study and measure the otherwise inaccessible structures and to watch the dauntless birds as, with unabated zeal, they set about preparing new cradles for their nestlings.

The female oropendolas built with no help from the much bigger males. Through much of the day they worked with tireless industry, although, like most birds, they built most actively early in the morning. Among the materials they used were long, pliant fibers ripped from leaves of banana plants in a neighboring field. I tried more than once to watch them at this activity, but a vigilant male nearly always accompanied a small party of females on their excursions to gather material and stood like a sentry in a coconut palm or in some other commanding situation. His loud *cack* of alarm nearly always put them to flight before I could see as much of the process of stripping fibers as I desired. However, with persistence I learned how this was done. Standing on the massive midrib of a huge banana leaf, the oropendola, taking advantage of a transverse tear made by the wind in the broad blade, bent down and nicked the smooth lower surface of the midrib with her sharp, orange-tipped bill, then pulled off a thin strand, sometimes as much as 2 feet (61 centimeters) long, from the fibrous outer layer. Then she doubled her harvest in her bill and started toward the nest tree with labored flight, long strands streaming behind her.

On reaching her destination, the oropendola might rest briefly before she went to a fork or, less often, a simple twig, always at the outside of the tree's crown, where she wrapped and knotted the fibers to form a firm and stable attachment. When this anchorage for the future nest was sufficiently strong, the bird forced apart the tangled strands with her whole body until they made a loop in which she could rest while she continued to weave. Now, on each return with more fibers, she perched in the ring, always facing the center of the tree, until a lengthening sleeve was formed in which she hung, head downward, while she wove at its lower margin. Presently the nest became so long that only her yellow tail was visible projecting through the doorway at the top, and at intervals her head appeared at the open lower end. Finally, her sleeve was long enough to contain the whole oropendola. After she had woven in the rounded bottom, I could catch, through my binocular, only an occasional glimpse of the orange tip of her bill protruding from the side of the nest as it pushed and pulled additional fibers through the closely woven fabric. As long as the bottom of the nest remained open, the building oropendola might most easily have emerged through it, but she always turned around and climbed up to leave through the top. Thus, the loop with which the pouch began became the permanent doorway.

In addition to fibers from banana plants, the oropendolas gathered for their nests many slender vines with leaves attached and long, narrow strips torn from the fronds of palms. Such was the patient application of the birds that these great swinging pouches, from 2 to 5 feet long and from 7 to 9 inches in diameter near the bottom (60 to 152 by 18 to 23 centimeters), were completed in an average time of ten days. One bird, who appeared very hurried, finished a somewhat shorter nest in seven or eight days, while another, who blundered much as she started her nest, took twice as long. Probably these replacement nests were built more rapidly than nests begun early in the season. Chapman (1929) found that, as nesting began, Chestnut-headed Oropendolas took three or four weeks to weave their usually shorter pouches.

Fully to appreciate the quality of the oropendolas' weaving, one should cut open a fallen nest and spread it to the light. It forms a regular, even fabric, with meshes wide enough to admit air to the brooding female and her nestlings yet strong and durable.

The construction of these nests did not proceed without frequent bickering among the laboring female oropendolas. Often two nests were started so close together that the builders, getting in one another's way while they wove, paused to express annoyance in loud, high-pitched, irritated voices, like children who interfere with each other at play. Sometimes each menaced the other with open bill or, completely losing their temper, they faced one another in the air as they fluttered downward until their proximity to the foliage below warned them that it was time to cease quarreling. Then they separated and flew up to continue to weave side by side as though they had never disagreed.

What most surprised me in these usually orderly, industrious birds was how frequently they stole materials. A building oropendola could rarely resist the temptation to snatch a fiber that hung loosely from the unfinished nest of a neighbor to incorporate it into her own. Sometimes, when the upper end of such a strand was attached more securely than she reckoned, the would-be thief, grasping it firmly in her bill, hung with half-closed wings beneath the nest until the coveted piece came loose or the owner returned to drive her away. Foiled at one attempt at theft, an oropendola might go straightway to steal from another nest. In the long run, I believe that this habit of pilfering must benefit the colony, for it discourages careless construction. It is not easy to extract a fiber that has been well woven into the fabric of a pouch. Individuals who build most carefully and leave fewest loose strands are least often troubled by thieving neighbors, and they finish the stronger nests.

The effrontery of these robbers did not end here. Often, returning with long banana fibers trailing behind her, an oropendola paused to catch her breath on a neighboring bough before going to continue weaving her nest. Seeing these strands depending from the bird's bill, a second female, perhaps just starting to go afield to gather material, would change her course and grab one dangling extremity, then, hanging from it with nearly closed wings, do her best to pull it away. The utterance of one single note of protest would have been disastrous to the rightful owner, for to open her mouth, as did the raven with the cheese in the fable when he succumbed to the flattery of the fox and began to sing, would have resulted in the loss of everything. The situation was so ludicrous that at times I could hardly avoid laughing aloud, at the risk of alarming the birds. Again, two oropendolas perching side by side on a bough would engage in a tug-of-war over a billful of fibers and perhaps in the end unwillingly divide the spoils. Hens from whose nests or bills material was pilfered usually took their losses without showing emotion; if fights arose from this cause, they were less frequent and prolonged than those over nest sites. But on the whole the female oropendolas were too busy to waste much time pilfering or quarreling, and weaving proceeded apace.

One morning, when the female of nest number twenty-two was away gathering material, another came up to her unguarded nest, which was then in the loop stage, and tore at it vigorously, pulling apart the fibers and destroying the neat appearance of the work. The loose ends left by the first thief attracted others who made repeated visits and continued the undoing of the structure until the loop was nearly torn apart at the bottom, where it was held together by only a few strands. The builder's efforts to repair the damage were neither very intelligent nor very successful. While working, she grasped with a foot each arm of the loop and straddled the gap in what appeared to be a most uncomfortable position. She continued to knot fibers along the sides of the loop without at first succeeding in closing the bottom, and while she struggled with the situation her neighbors persisted in stealing from her. For two days the loop lengthened without becoming sound. Finally, however, the oropendola succeeded in bridging the gap and completed a serviceable nest, which differed from most chiefly in its longer entrance.

Watching the efforts of this oropendola and of another who had much difficulty forming her loop on a straight twig convinced me that these birds, skillful artisans though they are, are deficient in mechanical sense. Both of these females had sufficient skill in weaving to have repaired their nests promptly, but they failed to grasp at first what needed to be done. A few strands carried across the gap and knotted at either side, with knots that these birds knew well enough how to tie, would have solved their difficulties in an hour, but they blundered along for several days before they managed to mend the breaks.

The male oropendolas, as big as crows, were a third larger than the females, whom otherwise they rather closely resembled. Since they were much less numerous than females, polygamy necessarily prevailed in the colony. At the nest tree, the males gave no indication of being mated to particular hens or to groups of them; they were ignored by the latter and mostly ignored them. While the females labored assiduously at their nests, the big males strutted about the tree with heads held high and pompous gaits. Although idle, they never quarreled among themselves. Sometimes, indeed, one male darted at another who promptly retired and thereby avoided an encounter, since the pursuer always quickly forgot whatever cause of enmity he might have had. At intervals, these males voiced their far-reaching calls. Bowing profoundly, until his raised tail stood directly above his inverted head, lifting his spread wings above his back and ruffling all his body feathers, the oropendola would utter, or rather seem to eject with great effort, an indescribable liquid gurgle. Heard from afar, there is no sound, save possibly the ventriloquial call of the Short-billed Pigeon or the mellow notes of the Great Tinamou, which to me is more expressive of the mystery and wonder of the rain forests of tropical America; but close at hand the effect of the oropendola's call is somewhat marred by screeching overtones, as though the machinery that produces this inimitable sound were in need of lubrication. These bows and gurgles were not addressed to individual hens so much as to the world at large.

The males were the watchmen of the flock. At the approach of danger, real or fancied, and not infrequently with no evident cause of alarm, they voiced a sharp, harsh *cack* which sometimes sent the whole flock dashing headlong into the nearest sheltering thicket but at other times was ignored by most of the community. The approach of a man was usually greeted by a few such notes of alarm; but if the colony had not been persecuted, few or none of the building hens heeded the warning; and thenceforth the man might stand quietly in full view and watch all the activities of the colony without causing the least unrest. One or more male oropendolas usually accompanied each party that left the nest tree to forage or procure building materials, and it was extremely difficult to elude the keen eyes of these sentries. The birds were as shy away from the nest tree as they were bold and confident among its boughs; when they were encountered afield, a male's shout of alarm invariably sent them into instantaneous retreat.

After they finished weaving their pouches, the female oropendolas neglected them for a day or two when courtship and mating probably occurred off in the forest (Howell 1964). After this interlude, the hens returned and worked diligently for three to six days more, plucking dying or dead leaves from trees at a distance, tearing them between foot and bill into pieces an inch or two long, and carrying them into their pouches. These fragments formed a yielding litter in the bottom and apparently served to keep the two eggs from rolling together and breaking when a strong wind rocked the swinging basket. Sometimes at first a bird brought leaves and fibers alternately, as though she had started to line the nest before she had quite finished weaving. Even while incubation was in progress or while the nest held young, a female occasionally took pieces of leaves into it, more rarely a fiber. After the completion of the new nests, the colony contained a total of eighty-eight, but not all were occupied.

A bustling activity prevailed in the nest tree at the height of the breeding season in May: the female oropendolas hurrying back and forth with fibers or food for the young in their bills or clucking as they paused to rest among the branches; their not infrequent quarrels and high-pitched remonstrances; the male birds strutting along the limbs or flying with resonant wingbeats from one to another, often repeating their liquid calls; the whining cries of nestlings rocked in their lofty cradles; the voices and movements of birds of many kinds that came to the nest tree to hunt for insects or to rest.

This throbbing activity was augmented by the disturbances caused by lurking Giant Cowbirds. So long as any nests remained in which incubation had not yet begun, the big, black female cowbirds continued to haunt the nest tree. Of all birds, save of course dangerous raptors, their presence alone was resented by the oropendolas. Continually chased by first one and then another of the larger yellowtails, the cowbirds circled around and returned with undaunted persistence, entirely unabashed by repeated rebuffs. By skulking most of the day in the nest tree and frequently looking into the openings of the nests, they apparently kept themselves well informed as to the condition of each, and at the proper time they tried to slip in and lay their eggs. Sometimes they miscalculated the hour of their visit and attempted to enter a nest while the oropendola was at home; but they beat a hurried retreat when the irate owner came out as though to demand an explanation of the intrusion. Time and again they were driven away at the very point of entering a nest; but, never despairing, one would finally elude many watchful eyes and slip into a nest, where apparently she left an egg to be incubated by the female oropendola.

The Giant Cowbird who lays in an oropendola's nest meets far more opposition than parasitic birds of other species who deposit their eggs in nests of small birds that breed in isolated pairs. The Giant Cowbird that finally succeeds in laying in an oropendola's pouch must not only dodge the watchful oropendolas of both sexes but sometimes also outwit jealous rivals of her own kind, each eager to drop her own eggs into the newly finished nest and ready to drive away another cowbird who attempts to get ahead of her. It costs the cowbirds so much effort to foist their eggs on oropendolas that I suspect it would involve very little more labor to build some simple nest and rear their own young.

Had the oropendolas made a concerted

attack upon these unbidden guests they might have driven them permanently from the colony, but they are mild-mannered birds who seemed to be satisfied if they could prevent the entry of the cowbirds into their nests. They rarely continued the pursuit of a cowbird for more than 5 or 10 yards (4.5 or 9 meters) from the doorway of a nest and never farther than the next tree. Long-suffering as they were, they were not always deceived by the foreign eggs in their nests. Early one morning toward the end of May, several cowbirds flew into the nest tree and were unusually persistent in trying to enter some of the nests in the group most recently completed. Several times one was on the point of slipping into a nest in which the owner was sitting, but always the owner hurried out and began with great spirit to pursue the black intruder, who fled uttering harsh, nasal whistles. Soon there were several cowbirds chased by as many oropendolas, and all was confusion in that part of the tree. Apparently, one cowbird took advantage of the disturbance to enter a nest and leave an egg within. What I actually saw a few minutes later was an oropendola emerge from one of these nests with something white in her bill. Flying to a higher branch, she dropped what she held. With a loud *plunk* it struck the leaf of a banana plant growing in the thicket beneath the nest tree, rolled off, and came to rest on the dead leaves that carpeted the ground. When I picked up the egg, it was unbroken by its 80-foot (24-meter) fall and still warm. It was white with a tinge of blue and marked with a few scattered, inconspicuous scratches of brown. Fortunately for the continued existence of cowbirds, other oropendolas are more easily deceived than this particular hen, and sometimes, as I have seen, they rear the offspring of the Giant Cowbird.

After incubation began, I could only surmise how things were proceeding inside the closely woven nest pouches.

Often an idle male flew with sonorous wingbeats to the side of a nest, upon which he pounced heavily. One seeing him do this for the first time might fear that he had torn it, especially since at the moment of making contact with the pouch one heard a sound that suggested the ripping of strong linen. This, however, was a vocal utterance that I often heard from males at a distance from the nest tree. Clinging head downward to the side of the pouch, the big bird would lift his wings above his back and emit his liquid gurgle while the hen brooding within answered with a high-pitched whine, in which she was sometimes joined by the entire group of birds whose nests were jarred by his boisterous behavior.

One day a Mica, a thick black-and-yellow snake fully 6 feet (2 meters) long, climbed up the tangled mass of vines that reached halfway up the bole of the tall nest tree. When the snake reached the top of the vines, it was spied by a pair of Great Kiskadee Flycatchers whose nestlings reposed in a domed nest in the crotch of another tree at least 30 feet (9 meters) away. These bold birds could not tolerate an enemy of such formidable size so close to their offspring. They darted toward the snake's head, often shooting past hardly an inch from its nose, voicing an angry *eee* when nearest it. A female oropendola descended to this lower level to see what the excitement was about but did not join in the attack upon the snake, and two inquisitive cowbirds perched dangerously close to the reptile. A few courageous lunges by the kiskadees stopped the advance of their enemy, which turned and went slowly downward among the vines; but until it had vanished into the rank vegetation near the ground the flycatchers continued to menace it. I believe that even without the kiskadees' threats the length of clean trunk between the top of the embracing vines and the lowest limb of the oropendolas' tree would have proved a barrier insurmountable by the Mica. A few years later, an experience at a colony of Yellow-rumped Caciques demonstrated how devastating the intrusion of such a serpent can be (Skutch 1979).

The ever-active oropendolas continued their daily tasks even after the sun had sunk behind the western mountain rim. As dusk descended over the quiet valley, they wove a last strand into an unfinished nest or carried a final mouthful of food to waiting nestlings. Others sat among the high boughs of the nest tree, enjoying evening's refreshing coolness, the hens clucking in contented tones, the males often bowing as they uttered their stirring songs. Then, in little groups of several together, the males, and females who had neither eggs to incubate nor nestlings to brood, flew northward, whence floated the calls, mellowed by distance, of males who had preceded them. Finally, as the tree frogs began their shrilling and the soft *kwahreo* of the Pauraque announced the end of the day, the last lone male oropendola jumped heavily to the side of a nest, sang loudly while making his deep bow, then flew with resounding wingbeats to overtake his companions. Meanwhile, the remaining hens retired one by one to cover their eggs or nestlings in the swinging pouches. Older nestlings slept alone until their mothers returned with food at dawn. Every oropendola who remained in the tree was in a nest.

At daybreak the adults returned to their nest tree, flying up the valley with wingbeats as regular as a crow's and appearing as black in the dim early dawn. Mothers of nestlings stopped along the way to find food for them and arrived at the nests with the first installment of breakfast in their bills.

How long a female oropendola incubated her eggs I could not tell, but the period was probably about two weeks. Finally came a day when a female was seen returning to her nest with fruit that she had gathered at some distant tree, assuring me that her nestlings had hatched. Soon their quavering nasal whines were sufficient evidence of what the nest held. This was a busy time for the female oropendola, for, as she had built her nest and incubated her eggs alone, so she fed her nestlings unaided by the males, and the food must often be procured at a great distance. These big birds, nesting in populous colonies, can hardly find enough food near their nests in the manner of smaller birds that nest alone. Returning with food in her bill, the hen flew directly toward the nest's entrance and, when almost there, folded her wings and glided deftly in, never pausing until the pouch had swallowed up her form. She seemed aware that to delay with only her head inside the doorway would needlessly expose her to the attack of a hawk whose approach she could not see. Emerging was different; often she delayed with the posterior half of her body still in the nest while she surveyed the outer world before launching forth.

By noting how long an oropendola continued to carry food into a nest, I learned that nestlings remained in their lofty cradles about thirty days, a period unusually long for a passerine bird. By the time the occupants were ready to fly, nests showed many signs of their long use. Newly completed nests were more nearly cylindrical than old ones, which became strongly pyriform as a consequence of the continued weight in the bottom. The female's innumerable passages through the entrance often tore it downward until it extended far along one side of the pouch. Before the nestlings left, several nests hung precariously by only a few strands and swung dizzily as their mothers went in and out while attending them. A few nests, torn from the supporting twigs, did not fall because, while building, two females had unintentionally sewn together pouches that hung in contact, and the stronger upheld the weaker. The birds made no attempt to strengthen these inadequately attached nests, as they might have done by lacing the attachment with additional fibers. The males quite failed to understand the delicacy of the situation, for they continued to jump on these damaged nests, just as they did on the sound ones, to call while hanging head downward. The repeated impacts of these heavy birds jeopardized the nestlings and in some cases hastened their fall. The males often turned their heads sideward to peer with one eye into pouches that sheltered nestlings, but despite their evident interest they never entered. It did not seem to occur to them that these nestlings might be fed.

Early one misty morning toward the end of May, I found the hens flying about the nest tree, clucking earnestly, possibly to coax from its nest a fledgling ready to leave. Soon I saw a pouch bow down at short intervals, as though something fairly heavy were moving upward inside by intermittent advances. Presently the pale head of a young oropendola appeared in the doorway. Perhaps startled by its first view of the outside world from so great a height, the young bird paused, seeming uncertain what to do. Its mother, who had been circling around the tree, now passed close in front of the nest, clucking as though to encourage the fledgling, who now flew boldly forth. Without touching a branch of the nest tree, it followed her across the rivulet to a small tree on the hillside to the west, completing a flight of about 200 feet (61 meters) on a descending course—not a bad achievement for a young bird who had never before been able to spread its wings fully! Several other hens who had followed the fledgling on its first flight now joined its mother in driving away a Brown Jay who had entered the tree where it perched. After feeding the fledgling here, the parent coaxed it still farther from the nest tree; her every effort seemed to be directed toward leading it to a distance from the colony.

Another newly emerged fledgling, whose mother was also feeding a sibling still in the pouch, delayed among the nests for at least two hours, but by the end of the morning it had vanished. It was most exceptional to see an exposed fledgling in or near the colony; most appeared to have been led away as soon as they abandoned their nursery, like the one whose departure I witnessed. The nest tree was too conspicuous and widely known for the young to linger there with safety after they had left the protection of the pouch. Even the adults did not pass the night in so exposed a situation.

The last young bird to depart the nest that year was, to judge by its size, a male. For a full month I had watched his mother carry food into the pouch. Evidently, his development had been somehow retarded; when at last he emerged he could not fly like the youngster whose departure I had watched earlier but fluttered down into the thicket beneath the nest tree, where, despite his ineffectual attempts to flap away, I caught him without much difficulty. When I was about to pick him up, he beat his wings frantically and squawked madly, while his mother and half a dozen other oropendolas flew around above us in the greatest excitement. When I offered my young captive a finger as a perch, his long toes closed on it with such an iron grip that I winced.

Never have I had such an escort of birds as that which followed us back to the house. The one who seemed to be my captive's mother led, flying along from tree to tree and keeping very close to us, clucking with all her might and uttering a kind of scream that I had not previously heard. When we reached the house, the other oropendolas dropped away; but the mother remained with us, perching high in a neighboring tree with an orange-colored fruit in her bill while we photographed her offspring. To make sure that our portraits were adequate, we kept the young oropendola on the screened porch of the building while we developed the film. All this while, the parent flew around the house in great circles, answering the juvenile's calls. Finally, we returned the impatient fledgling to the spot whence I had taken him a few hours earlier. Here his mother found him, and in this same thicket she attended him during the next three days, after which he disappeared. The whole episode was a display of parental devotion unique in my experience with birds.

After the departure of the last young oropendola, the colony in the treetop stood like a deserted village. Already some of the nests earliest built had fallen from age and decay. Although it was only the end of June, eight or ten of the newly completed nests were abandoned before I saw any food for nestlings carried into them. The young in some neighboring nests had fallen victims to an unknown predator who might have taken the eggs from these. Each of a number of fallen nests picked up beneath another colony in the same region had a hole a few inches in diameter in the wall near the bottom; of unknown origin, they had been made by cutting and forcing apart the fibers. Meanwhile, successfully fledged young of the year followed adults of both sexes up and down the valley in flocks of usually less than a dozen individuals. For many weeks after leaving the nest, the young were fed by their mothers, who gave food to immature males much bigger than themselves.

Strangely, I never saw a young Giant Cowbird in these flocks. But two years later, in the Motagua Valley of Guatemala, I watched a solitary female feed a cowbird who cried with nasal notes and shook its relaxed wings like a fledgling. Then she flew over a banana plantation with her black protégé following. Apparently, the oropendolas will not tolerate these intruders in their flocks and the foster mother leads a lonely life rather than abandon her fosterling.

The male oropendolas' interest in the nests continues after the close of the breeding season. Long after the last fledgling has flown from the nest tree, males from time to time return to its crown and examine the deserted nests, perching above a doorway and peering down into the pouch as they do while young are within. At other times they jump to the side of a nest, with the peculiar sound of ripping cloth as they alight, and hang head downward to deliver their songs. In the last months of the year, when the females usually associate in flocks, one often sees a lone male perching at the top of a lofty tree where, scarcely bowing, he pours forth his stirring, liquid call.

Montezuma Oropendolas are found throughout the rainy eastern lowlands of Middle America from northeastern Mexico to the neighborhood of the Panama Canal, where they have recently arrived. Where the continental divide is low, as in Honduras and northern Costa Rica, they spill over to the Pacific slope, but they avoid the drier parts of the Pacific lowlands. From the coasts they range upward to about 4,500 feet (1,370 meters) above sea level and rarely 1,000 feet (305 meters) higher. Since for safe nesting they need a tree with a tall, clean trunk that stands apart from its neighbors, its crown difficult for flightless predators to reach, their colonies are more often in clearings than in unbroken forest. Sometimes their nest tree is close to a building, a railroad, or a highway. If a sufficiently large tree is not available, they may distribute their pouches over several smaller ones. Seven thorny Pejibaye palms, growing close together in a pasture, supported a total of sixty-one nests, from one to twenty-two in a palm. These nests were attached to the rachises of the big, feathery fronds, well out from the trunk, with sometimes three or four of the long pouches hanging in contact.

When not nesting, Montezuma Oropendolas travel in large, straggling flocks, sometimes in company with smaller Chestnut-headed Oropendolas. Each day they cover long distances in their search for food, perhaps more often in clearings than in heavy forest, where they gather fruits in the crowns of the great trees. They eat the green fruiting spikes of Cecropia trees and the small seeds, enclosed in red arils, of the tree *Alchornea costaricensis*. They frequent neglected plantations, where they feast upon ripe bananas and suck nectar from the staminate flowers, clinging head downward from the knobby stem of the inflorescence while they insert their bills into the white flowers clustered beneath the fleshy, red, upturned bracts. Then they raise their heads to permit the sweet liquid to flow into their throats, as when they drink water. Likewise, they extract nectar from blossoms of the Orange Poró, a South American tree widely planted to shade coffee plantations in Costa Rica. When they drink from the red flowers of the introduced Flame-of-the-Forest tree, it is uncertain whether they are attracted by nectar, rainwater that has collected in the upturned, trumpet-shaped corollas, or a combination of both. They are more clearly seeking water when they split the big, furry, unopened calyxes filled with secreted liquid in which the inner organs of the flower develop, protected from the attacks of insects.

As the day ends, the Montezuma Oropendolas may seek a dense stand of tall, introduced timber bamboos, where with much hubbub they settle down to roost with Chestnut-headed Oropendolas and Garden Thrushes. But they are much more easily disturbed at their roost than at their nest tree, and if closely watched as they retire they may move to a more sequestered site.

11. *A Perennial Aspiration*

A few generations ago, men of European origin, especially in the New World, were exploiting the vegetable and animal kingdoms with little regard for anything except their own profit and pleasure. They would have been impatient of any attempt to moderate their lavish exploitation of nature's bounty. Now, intelligent people almost everywhere are awaking to the fact that such reckless exploitation cannot continue without preparing great hardships for future generations, if not for people now alive, for it is undermining the very foundations of human life. Today, despite the widespread persistence of short-sighted selfishness, it is not difficult to enlist support for any sound regulatory measure urged in the name of the conservation of natural resources.

Although no thoughtful person will deny the necessity of conserving the natural resources that support human life, this objective fails to go far enough to satisfy an awakened conscience. Beyond the question of whether we are preparing shortages and famines for future years is that of how our acts affect living creatures themselves, viewed not as sources of food or other necessary products, nor even as objects of scientific interest and aesthetic appeal, but as beings akin to us which should be regarded as ends in themselves and not merely as means to human ends. We view our relations with nonhuman creatures as constituting not simply an economic problem but also an ethical problem. We are certain that our treatment of animals and plants falls properly within the purview of ethics because, in addition to affecting the welfare of living things, it subtly affects the character of the actor, as is true of every act that has moral overtones. Kind and generous treatment of nonhuman creatures exalts and expands the spirit; callousness or cruelty to them, resulting from a defect of character, causes its further deterioration. When we approach nature ethically, we are interested not only in how our treat-

85

Agouti

ment of living things affects them, considered as ends in themselves, but in what it does to the human spirit.

Although recently it has been spreading, this attitude toward the rest of the living world is still somewhat exceptional in Western civilization. Those who adopt it may find themselves alone, without the moral support that many people need in order to persevere in their best resolutions in moments of doubt and weakness. We might be told that we are inordinately soft and sentimental, that our attitude is irrational or incongruous in a world so full of strife and bloodshed as this. In our isolation, we may even come to suspect that we are not made as other people but are somehow abnormal or queer. When we cannot find support for our ideals among those who immediately surround us, we do well to seek it at a distance; if we search diligently, we may find more than we imagined could exist. Indeed, if we pursue our investigations with care, we may conclude that it is not those who seek to regulate their dealings with nonhuman creatures by ethical principles who are exceptional, but rather that Western civilization is somewhat peculiar. Few cultures of the past have taken such a lax attitude toward man's treatment of animals; in few have religion and ethics done so little to mitigate his impact upon other forms of life—a consequence, apparently, of the circumstance that Christianity is largely an urban growth, nurtured in the larger cities of the Roman Empire.

To expand our ethics beyond the narrow confines of the human race, to live in harmony with all beings and injure none, is an aspiration very ancient and widespread among men. It has roots in certain attitudes and practices of primitive peoples, and it comes clearly to the surface in some of the earliest civilizations. It is doubtless not just a coincidence that Jainism, which rather convincingly claims to be the world's oldest living religion, is of all religions that which most insists upon harmlessness to all creatures. It is related that Nemi Natha, who lived in India probably in the second millennium B.C., was so filled with compassion for the animals that were being led to slaughter for his marriage celebration that he suddenly left the wedding procession, renounced the world, achieved enlightenment, and became the twenty-second Tirthankara, or great teacher, of the Jains. We need not question whether this story is sober history or legend; either way, it vividly portrays a cherished ideal in the absence of which it would be neither told nor remembered.

The bedrock of Jainism, on which its whole ethical system is built, is the principle of *ahimsa*, or harmlessness to all creatures. Although the word is negative in form, *ahimsa* is not a passive attitude but an expression of compassionate concern for all beings, from the greatest of them to the least speck of sentient dust. In the life of devout Jains, whose community still flourishes in India, it leads not only to constant care and strenuous effort to avoid harming the humblest of living things but to active beneficence. The Jaina monk tries to avoid careless movements that may crush some small creature that he fails to see. Jaina farmers protect their crops by scaring away rather than killing the free animals that come to feast upon them, while householders carefully carry troublesome creatures from their homes rather than crush or poison them. In celebrating their joyous occasions, Jains distribute food to animals no less than to people. Since the Jaina rules of conduct must be practiced in thought and word no less than in deed, even to think of killing any creature or to wish that harm befall it is a transgression. What this implies in a warm country, with its teeming and often annoying life, only one who has dwelt in the tropics can appreciate. A Jaina writer (Chakravarthy 1954) pointed out that there is no religious significance in obeying the injunction "Thou shalt not kill" when, as in the West, its application is limited to human beings, because we are restrained from murder by statute laws that severely punish the offender. Only when extended to creatures not protected by the police does abstention from killing acquire spiritual significance.

The doctrine of universal harmlessness was accepted by the founder of Buddhism and finally passed over to Hinduism, where it helped abolish the animal sacrifices so prominent in the old Vedic religion, although not in all sects. The Buddhist commandment not to kill, unlike the corresponding Mosaic rule, does not refer only to men but is of universal application. This tenderness toward living things extended not only to animals but even to plants. From its cradle in northeastern India, Buddhism, the world's first great missionary religion, spread far and wide through Asia and the Eastern Archipelago, everywhere teaching the people gentleness and restraint in dealing with the living world and doing much to preserve the wonderful fauna of the oriental tropics in the face of dense human populations. Even on the bleak plateau of Tibet, where the harsher conditions of life discouraged the vegetarianism usually associated with Buddhism, travelers were impressed by the tameness of large free mammals and birds on the very outskirts of the principal cities before the Chinese conquest of the country.

When, a few centuries after the death of its founder, Buddhism crossed the Himalayas into China, it found that Taoism, despite different metaphysical postulates, had already done much to make the people gentle and considerate of all living things. It is recorded that T'ao Hung-ching, a Taoist physician, was debarred from heaven for having included such animate beings as leeches and gadflies in his materia medica. After he had prepared a new edition, substituting vegetable for animal remedies, he was admitted to bliss (Giles 1948). In the *T'ai-shang kan-ying p'ien*, an ethical treatise inspired by Taoist principles, we are admonished to turn toward all creatures with a compassionate heart. Even the multifarious insects, herbs, and trees should not be injured. A man is praised for having liberated captive birds and refrained from killing animals. But in China compassion for animals was not, before the introduction of Buddhism, confined to the Taoists, although most strongly stressed by them. In the fourth century B.C. the sage Mencius, as a Confucian a member of a formalist school often at odds with the intuitive Taoists, asserted that the higher type of man is loving to his parents, benevolent to the people, and kind to animals.

In ancient Persia, Iranian dualism led to an attitude toward living things that contrasted sharply with the unrestricted benevolence of the Jains, the Buddhists, and the Taoists. Among the Zoroastrians, some animals, including cattle, dogs, hedgehogs, otters, and certain birds, were held to have been created by the good Ahura Mazda; it was a religious duty to treat them with kindness and mercy. Terrific punishments awaited the person who slew a river otter. But ants, scorpions, frogs, snakes, and many other troublesome animals as well as some noxious plants belonged to the evil creation of the wicked lord Ahriman; to destroy them was a meritorious act. Unlike Dante in his journey through hell, his Zoroastrian counterpart, Arda Viraf, found souls in torment for mistreating animals no less than for sinning against humans, although some mitigation was allowed for the laudable activity of destroying creatures that Ahriman had made (Haug and West 1872).

Although apparently of later origin than the absolute *ahimsa* of Jainism, the Zoroastrian attitude toward the living world is more natural and spontaneous. Unless otherwise taught, children tend to assume a protective attitude toward animals that ingratiate themselves by their beauty, friendliness, and innocence, but they show scant mercy to such as inspire repugnance or dread by their strange or grotesque forms, menacing postures, or actual ferocity. They cherish birds and butterflies but kill snakes and spiders. Only with growing insight do we realize that it is impossible to divide the living world into the creation of Ahura Mazda and the creation of Ahriman in this naïve fashion. Good and evil are otherwise mixed on this planet, so that every creature from the best to the worst appears to have some share of both; and it is difficult to point to one that is wholly good, in the sense that it is harmonious in all its relations and injures nothing, or yet to one that is completely evil.

Despite the unhappy consequences of Zarathustra's teaching for certain kinds of animals, we are grateful to him for wrestling bravely with the baffling problem of evil and indicating one of the possible solutions. The monotheism of Israel avoided this pitfall of Zoroastrian dualism. The Messianic vision of the Hebrew prophets included peace and joy for animals no less than for men. Isaiah 11:0–9 proclaimed: "The wolf also shall dwell with the lamb, and the leopard shall lie down with the kid . . . and the lion shall eat straw like the ox . . . and the suckling child shall play on the hole of the asp . . . They shall not hurt nor destroy in all my holy mountain."

Unlike the Jains and the Buddhists, who believed that only by man's abnegation could strife be abated, the Jews have been content to await a miraculous advent and have denied themselves little to further the realization of the Messianic ideal as expressed by one of their prophets of widest sympathies. Yet their ancient code was not devoid of provisions for the welfare of animals both domestic and free: work animals were to enjoy the Sabbath rest with their masters; it was a duty to help an overburdened ass that had fallen, even if it belonged to an enemy; and it was unlawful to take a parent bird along with its eggs or young. The Apocrypha and Pseudepigrapha of the Old Testament and even more the Talmud emphasize again and again the duty of kindness to all animals. The rabbinical writers condemned hunting and attendance at the circus, with its bloody spectacles, as a surrender to one's lower passions and taught that a man should feed his work animals before he sat down to his own meal.

When we consider ancient Egypt's many restrictions upon the slaughter of animals, it is difficult to disentangle the influence of the spirit of kindness from that of the stubborn conservatism that carried primitive totemic cults well into the historic period. At least some of the priests would kill no living thing except the animals they sacrificed to the gods. Each district had its sacred animal which was not only strictly protected but sometimes cosseted to what seems to us an absurd degree. In Plutarch's day, the sheep was exempt from slaughter everywhere except among the Lycopolitans, who in eating mutton imitated their sacred animal the wolf.

Although we generally associate the doctrine of universal harmlessness with India, it was by no means absent from Greco-Roman civilization. Pythagoras early introduced among his disciples respect for all life, not unlike that which at the same epoch was becoming widespread in distant India and far surpassing in scope anything of which we have knowledge in the intervening lands. The ancient philosophers, even when, like the Epicureans, they denied that the gods took an active interest in the world, generally believed in their existence as blessed immortal beings who served as exemplars of what mortals aspired to be.

One of the attributes of the gods, in the Stoic no less than in the Epicurean and Neo-Platonic systems, was their harmlessness. In his *Epistulae Morales* the Roman statesman and philosopher Seneca, who amid the atrocious barbarity of Nero's reign preserved his faith that man is naturally the gentlest class of beings, wrote of the gods: "One who thinks that they are unwilling to do harm is wrong; they *cannot* do harm. They cannot receive nor inflict injury . . . The universal nature, all glorious and all beautiful, has rendered incapable of inflicting ill those whom it has removed from the danger of ill."

The Stoics did not carry their doctrine of harmlessness so far as to avoid killing animals for food, because, along with other Classical philosophers, they were interested more in the philosophic problem of justice than in the religious virtue of mercy, and not without reason they maintained that justice, in the strict sense, could be established only among rational beings. But the gentle and humane Plutarch, whose moral writings deserve to be as well known as his biographies, attempted to demonstrate, in several amusing essays, that the Stoics were wrong in denying that animals can reason, and he pointed out that to refrain from killing them for food would have been more consistent with their principles. Porphyry, the Neo-Platonist, recommended abstinence from flesh because God injures nothing; accordingly, the devout man who aspires to become godlike carefully avoids all injury. Yet in one point Plotinus' great disciple fell short of the Indians; he held that one may kill, although not eat, animals that endanger human life, whereas Jainism and Buddhism in their pure forms know no such reservation, at least for the members of their monastic orders, upon whom alone the rules of conduct were binding with full force.

When the Classical world was overrun by northern barbarians, this growing tenderness toward all living things, contrasting so sharply with the hideously brutal practices of the Roman arena, decayed along with so many other admirable aspects of the ancient culture. In medieval Europe, St. Francis of Assisi, who preached to the birds and the beasts, calling them his brothers, provided the outstanding example of the attitude we have been discussing, and his anniversary has been chosen as World Animal Day. Although exceptional in Europe, Francis' brotherliness with animals would hardly have won notice in India, where, apparently, it was routine practice at every hermitage to feed birds and other animals and water the plants as acts of piety. Kalidasa gives us a charming picture of this in *Sakuntala*.

While in the ancient East it was religion that chiefly inculcated harmlessness to all beings and in the Classical world it was philosophers imbued with religious fervor, in the modern West this aspiration has been most eloquently expressed by poets. Shelley, in the magnificent prelude to *Alastor*, claims the favor of the elements:

If no bright bird, insect, or gentle beast
I consciously have injured, but still loved
And cherished these my kindred . . .

The lesson of Coleridge's *Ancient Mariner* and Blake's *Auguries of Innocence*, with its accusing lines "A Robin Redbreast in a Cage / Puts all Heaven in a Rage," are too well known to English readers to need repetition here. Tennyson's *In Memoriam* voiced the hope

That not a worm is cloven in vain;
   That not a moth with vain desire
   Is shrivell'd in a fruitless fire,
Or but subserves another's gain . . .

In his *Ancient Sage* we find the injunction "Nor harm an adder thro' the lust for harm / Nor make a snail's horn shrink for wantonness." It would be easy to fill many pages with similar sentiments from poetry—and prose—in the English tongue, but these will suffice for our present purposes.

Before concluding this chapter, I cannot refrain from quoting a passage written by an American, the Quaker saint John Woolman, who wrote in his journal: "I . . . was early convinced . . . that as the Mind was moved, by an inward Principle, to love God as an invisible, incomprehensible Being, by the same Principle it was moved to love him in all his Manifestations in the visible World.— That, as by his Breath the Flame of Life was kindled in all animal sensible Creatures, to say we love God, and, at the same Time exercise Cruelty toward the least Creature, is a Contradiction in itself."

It is evident that, when we aspire to go through life harming no living thing, however much profit we might gain by it, we have not been captivated by some newfangled notion or one peculiar to ourselves and deserving the ridicule of hardheaded practical neighbors. On the contrary, we share an ideal that has been cherished for thousands of years in nearly every literate culture and by many of the wisest and best of men. Unfortunately, it is often difficult to learn what efforts the people who gave eloquent expression to this aspiration made to live in accordance with it. Perhaps it is better not to have ideals than to profess them yet take no steps to make them come true. Absence of ideals is a mark of a sluggish intelligence and lack of imagination; but to profess ideals and do nothing to advance them is a sign of moral lethargy and indifference, which is more pitiable than a dull imagination. No one who has seriously thought about the matter has ever supposed that the ideal of living in peace with all beings is easy to realize on a planet so packed with life as ours. But an ideal easy to attain is hardly worthy of us; a true ideal is a standing challenge to surpass ourselves, a goal that we can steadily approach, although it ever remains above our reach.

If we could synthesize the things that we use from widespread, abundant, simple materials, in the beneficent manner of plants, we might advance a long way toward the realization of the ideal of living in harmony with the life around us. Although, unfortunately, we cannot become plant-animals and absorb energy from sunlight, we can at least take a hint from vegetation. In broadest terms, what green plants do is to take simple substances and elaborate them into forms that are indispensable to life and frequently beautiful as well. Certainly in many fields men can do the same, thereby giving satisfying play to their creativity. The greater the difference in organization between the raw materials and the final products and the higher the uses to which these products are put, the more admirable this creativity becomes. We might call this the principle of maximum elaboration and elevation, which from another point of view might be called the principle of minimum destruction.

Let us begin by applying this principle to house construction. Many dwellings are made largely of wood. Vast expanses of temperate-zone forest have been leveled to meet the insatiable demand for timber, and now magnificent tropical forests, at great distances from the points where the lumber will be used, are feeling the heavy pressure of the market. Wood is already highly organized tissue of very complex structure, and the trees from which it comes are often so stately that only a very talented architect can design a building that rivals them in beauty. Accordingly, when a house is made of wood, it is questionable whether there has been any elevation of its chief material.

If, instead of wood, we build the house largely of cement, which is made of widespread rocks, we raise the level of organization of its materials. But even common clay can make a comfortable, attractive dwelling, forming not only the roof, of glazed or unglazed tiles, but likewise the walls, either as adobe blocks or more securely bound together in other modes of construction. Moreover, far less power is consumed in the preparation of clay than in the manufacture of cement. To raise clay from the subsoil to sheltering walls and even the roof over one's head is, both figuratively and literally, an excellent example of maximum elevation. Another advantage is that buildings properly made of minerals, such as clay or cement, tend to be more durable than wood, especially in warm, damp climates; diminishing the need of replacement, they still further reduce the drain upon products of the forests.

When men tear skins from animals to clothe themselves, they do not increase the organization of the materials but convert a living integument into a dead one. Since the skin was more necessary for the animal than for the human who wraps it loosely around himself, usefulness has hardly been increased. But to weave cotton or other vegetable fibers into cloth is an art so admirable that its invention in prehistoric times was attributed by the Greeks to the goddess of wisdom, Pallas-Athene. To cut and sew this cloth into a shapely garment elevates the original material a stage higher. Here again we have a great increase in organization and usefulness. With the continuous amazing advances in technology, it would not surprise us if before long clothing will be made from common earth.

The same principle may be applied to food. Since there is little difference in the level of complexity (other than of the brain) between other warm-blooded animals and man, to convert their tissues into human tissues hardly increases their organization or their usefulness to life as a whole. When we take cold-blooded animals into our bodies, the degree of elevation may be somewhat greater. It is very much greater when we nourish ourselves with the products of plants which, especially in the storage organs and fruits that are the parts chiefly used for human food, are far more simply organized than the bodies of vertebrate animals. The structure of grains is more elaborate than that of the edible parts of fruits, tubers, and storage roots, although it is still inferior to that of the higher animals or even of the plants that might grow from these seeds. In any case, the grains of cereals and other cultivated plants are produced in such great profusion that all could not find room to grow to maturity, so that if not eaten they would be wasted. The conversion of simple vegetable products into a human body is a very great elevation. Perhaps the highest degree of elevation that the living world as a whole presents is that which begins when green leaves bind the energy of sunlight in carbohydrate molecules and this energy finally blossoms forth in the discovery of truth, the production of beauty, or some noble deed.

The widespread application of this principle of maximum elevation would not only give gratifying play to human creativity but increase man's harmony with nature by greatly diminishing the drain upon its resources. Scarcely anything would so greatly advance the cause of conservation. In old, densely populated countries without much international trade, economic necessity, if nothing else, has forced people to comply with this principle rather closely: the houses, especially of the poor, contain a minimum of timber; clothing is of woven cloth; and the diet is largely vegetarian. In the East, the slaughter of animals is diminished by religions that inculcate compassion for all that lives. In the West, where abstention from animal flesh has in the past been practiced only by a small, sensitive minority, the rapidly growing concern for the environment may make vegetarians of people who are not distressed by what happens in the abattoir.

Among my friends is a young United States Peace Corps man in a tropical country, an ardent worker in the cause of conservation, who refuses to eat beef because his wide travels about the country have opened his eyes to the havoc wrought by cattle raising. The recent scarcity and high cost of meat in the North have greatly accelerated the conversion of tropical forests, both wet and semiarid, into grazing lands, with the destruction of the native vegetation and the many animals unable to survive in the radically altered habitat. Man alone is an alarmingly destructive animal but in association with grazing and browsing animals, including bovine cattle, sheep, and goats, he forms the most devastating team that this much-abused planet has seen. Over immense tracts of country, the native flora and fauna are swept away to make room for cattle. On steep slopes in rainy regions, the sharp hoofs of these heavy animals break the roots of herbage and plow up the soil, greatly accelerating erosion and leaching under heavy downpours. Not only does the habit of eating flesh greatly increase the amount of land needed to feed each person, but it hastens the degradation of much of this land. To the true friend of Earth there is no more depressing sight than that of whole mountainsides, which a few years ago were covered by splendid forests of great trees, reduced by grazing and erosion to barren slopes.

Although we need proteins to build and repair tissues, carbohydrates are adequate sources of energy and they are much more easily and cheaply produced in great quantities. To depend upon proteins for energy, as in a diet overrich in animal products, might be compared to breaking up fine furniture for firewood. As one becomes thoughtful, certain foods that once tasted good to the palate no longer taste good to the mind, and it becomes impossible to enjoy them.

Life, which began obscurely in primal seas billions of years ago, has by means of a long series of undirected mutations screened by implacable selection finally produced beings capable of cherishing the ideals that we have surveyed in this chapter and often, too, of trying strenuously to practice them. This is certainly one of the most wonderful outcomes of evolution, one which any evolutionary philosophy that pretends to completeness must take into account, no less than anatomical structures and physiological functions.

12. The Oriole-Blackbird

Among the birds new to me that I met among the hills south of Lake Valencia was a large, handsome species, about 11 inches (28 centimeters) long, golden-yellow on head, neck, and all underparts, black on back, rump, wings, and tail. Each dark eye was set amid bare black skin. Its sharp black bill resembled that of an oriole but was somewhat stouter. From the colored picture in Kathleen Deery de Phelps' useful little book, *Aves venezolanas: Cien de las más conocidas*, I readily learned that this gold-and-black bird is known as the Maicero, and its scientific name is *Gymnomystax mexicanus*. The second part of this name perpetuates a blunder not infrequently made by Europeans with hazy notions of the distribution of New World organisms that they described and named; widespread in northern South America, the Maicero is unknown in Mexico and Central America. In Meyer de Schauensee's *Guide to the Birds of South America*, this species is called the Oriole-Blackbird, from which I inferred that, although it resembles the orioles in plumage, in habits it is closer to the more terrestrial, largely black members of the oriole family, the Icteridae. I was eager to learn more about it.

For a long while after our arrival in Venezuela in March, I found these birds mostly perching high in trees growing in or beside pastures, often on the topmost exposed twigs, where they might have remained unnoticed if they had not called attention to themselves by a long-drawn-out nasal screech, ascending sharply toward the end, that sounded like the squeaking of a rusty gate and seemed unworthy of birds so splendidly attired. A frequent utterance which I took to be the birds' song was *chaa chaa chrick chaaa*, with the *chrick* clearer in tone than the nasal *chaa* and rising in pitch. Alike in plumage, the sexes differed little in voice, except that the male's songs tended to be longer than those of his mate.

Displaying to another Oriole-Blackbird who perched near him in a tree, one who was apparently a male raised his head, puffed out the feathers of his nape and back, fanned out his tail, and slightly

Oriole-Blackbird

spread and dropped his wings while he emitted the usual screech. While calling, these birds sometimes pointed their bills straight upward in a posture not unlike that of displaying male Great-tailed Grackles. At first, the Oriole-Blackbirds that I saw perching high in trees were often in trios, but as May advanced pairs became more common. Again and again, I watched hopefully for them to build; but after an interval of motionless perching high above me, with a *chrick chaa* they abruptly took wing, flying so high and far over the surrounding barren hills that I did not try to follow. How could I become familiar with birds that remained so exasperatingly aloof?

Finally, while I wandered over a wide pasture with scattered trees one morning in early June, a female with a straw in her bill flew into the top of a massive yagua palm. She disappeared between the bases of the younger of the great, feathery fronds of the palm's spreading crown, but presently she emerged, found more material on the ground, and carried it to the same site. Here, about 23 feet (7 meters) up, she was starting a nest.

Next morning, June 6, I arrived early to watch the birds build, but it was nearly seven o'clock before I heard their screechy notes or saw them. While the male Maicero perched conspicuously at the top of a neighboring tall dead tree or at the very apex of a more distant leafy one, voicing from time to time his long-drawn-out nasal screech, the female walked over the pasture where the new grass was sprouting, gathering pieces of dead herbage for her nest. Often she picked up, tested, and dropped a number of pieces before she found satisfactory material. Then, her black bill laden, she flew up to a frond of the palm and promptly vanished into the heart of the dense crown, where I could not see what she did. Usually she stayed out of sight for a good while before she emerged and flew down, often gliding on set wings and voicing a single sharp *cluck* or a longer *tuc-titit*, to gather more material from the ground. From time to time, her mate descended from his high lookout to pick up pieces of vegetation from the pasture and carry them to the nest site. He did this both while the female was present there and while she was on the ground, so that apparently he himself placed in the nest what he took to it. Once, however, he emerged from amid the palm fronds still bearing his material.

The female started to build at 7:10 A.M., and during the next hour she took material to the nest eleven times, the male three times. Most of these trips were made toward the end of the hour, while rain fell. From 8:10 to 8:38, the female carried material to the nest twelve times, the male only once. After this, building proceeded much more slowly. Much of the time the pair were out of sight, but once I watched them forage close together amid the wet grass. Soon after 9:00, they flew away and were absent so long that I left, too. In little more than two hours, the female had taken thirty billfuls of material to the nest, the male only five.

At 7:40 the next morning, I found the female at work. Now she was gathering what appeared to be mud or cow dung mixed with short fragments of vegetation. In the next quarter-hour, she carried five billfuls to the nest. Once the male came down from his high treetop and tugged at wiry stems that he could not tear loose. Then, apparently having found an ants' nest, he went through the movements of applying ants to his plumage—the only time that I saw a bird "anting" on the ground in the tropics, where this curious activity is nearly always done in trees or shrubs. Then he ate something. Finally, he gathered a billful of dead grass blades and carried them up to the palm tree, only to fly away still holding them. Again he took his billful to the crown of the palm but failed to deposit it on the nest. Instead, he bore it to a neighboring *Cordia* tree, where he and his mate went at intervals to eat the fruit.

Although I watched on subsequent mornings, I did not again see such sustained building as on June 6. One morning at about half past nine, I saw the female gather fine material from the ground to line her nest. After she had collected a generous billful, she started to pant in the bright sunshine, dropped her carefully selected load, and went to perch in the shade. She seemed most intolerant of heat. The last building activity that I saw was late in the morning of June 14, when the female took a large billful of the slender rachises of compound leaves to her nest. Thus, the construction of this nest continued for at least ten days. The work was done largely by the female, with her partner making occasional gestures of helpfulness.

While the Oriole-Blackbirds built, a pair of American Kestrels, or Sparrow Hawks, spent much time in surrounding trees, mostly ignored by the blackbirds. Once, while the female gathered material on the ground, a kestrel swooped down at her, but she avoided it by jumping into a bush that was conveniently near, then calmly resumed her task. While the male blackbird perched on a dead twig at the very top of a tall leafy tree, one of the kestrels flew at him, making him duck. When the raptor again dived at him, he moved to another branch a few yards away, and the aggressor alighted on that which he had left. For a while, the Oriole-Blackbird and the American Kestrel rested in the same treetop.

The completed nest was a broad shallow bowl, loosely constructed of materials merely piled together, with no weaving or interlacing. The foundation was of long, coarse straws mixed with smaller fragments of plants. The bulk of the nest was composed of the black, curving rachises, up to 6 inches (15 centimeters) long, from the compound leaves of an acacia-like tree that grew nearby. Although one morning I thought that I saw the female gather mud or cow dung—material used by grackles and other blackbirds—I found none in the nest after the young had gone. However, if only a little had been brought, it might have been washed away by the heavy rains of this season.

Since I could not, with available materials, construct a sufficiently long ladder so light that, unaided, I could raise it to the nest, I did not learn just when the eggs were laid. On June 16, for the first time I found the female on the nest late in the morning, suggesting that incubation had begun. Afterward, I enlisted a boy to climb a long pole into the crown of the palm tree and lower the eggs to me. There were three, light blue in color, marked, especially on the thicker end, with irregular blotches and speckles of black and lilac.

On June 21, I watched the nest in the palm tree from midday until the light grew dim at 6:45 P.M. Resuming my vigil the following dawn, I continued until 12:13. Only the female incubated. On June 21, she retired for the night as a shower began at 6:34, and she did not leave the next morning until 6:47, so that her nocturnal session lasted 12 hours and 13 minutes. On the two days, I timed ten sessions on the eggs, ranging from 17 minutes to 113 minutes in length and averaging 43.3 minutes. Ten recesses or absences from the eggs ranged from 8 to 35 minutes and averaged 23 minutes. For three-quarters of an hour during her longest session, in the late afternoon, rain fell with much wind. This session was followed by the blackbird's shortest absence, on returning from which she settled on her nest for the night. Her shortest session came in the middle of a warm morning. She incubated for 65 percent of her active day.

While the Oriole-Blackbird sat in her nest, I could at times glimpse the top of her golden head between the bases of the palm fronds, but otherwise she was invisible to me. On leaving, she often flew to the neighboring *Cordia* tree to eat the yellow fruits, after which she might descend to forage on the ground close to her mate. Often he accompanied her when she returned to the palm tree. She had no fixed route for reaching her nest: sometimes she would alight far out on a palm frond and walk inward to it, but at other times she flew right into the center of the tree's crown and promptly settled on her eggs. She rarely used the same approach twice in succession.

The male was most attentive. After escorting his partner back to the nest or near it, he would often rest in a neighboring treetop, preening and at intervals singing screechily. Sometimes the female left the eggs in response to his calls. From time to time, he visited the nest, either while the female was present or in her absence; but the obstructing bases of the palm fronds made it difficult for me to see just what he did there. He carried no visible food and apparently did not feed his incubating partner. One afternoon he stood for many minutes beside the nest, bending over his sitting mate, first from one side and then from another, while I vainly tried to see just what he was doing. Less than an hour later, he returned. This time, peering up between his legs, I could see him making movements which could have been preening her head or gently pecking her, perhaps to make her rise up and reveal the eggs. Next morning, he went to the nest just after his mate had left and doubtless enjoyed a satisfactory view of its contents, for this time his visit was brief. Once, while the female incubated, the male foraged through the tall pasture grass in close company with a flock of Smooth-billed Anis. While incubation was in progress, from two to four American Kestrels perched often in the trees nearest the palm that held the nest, but they were ignored by the blackbirds.

Between July 3 and 5, the nestlings hatched. Their yellowish flesh-color skin bore long but sparse gray natal down of the usual passerine type. The interior of their mouths was purplish red, and at the corners were prominent whitish flanges, about 2 millimeters broad. While my climber was in the crown of the palm tree, lowering a nestling to me, the two parents circled so close above him, uttering sharp *chip*'s and angry, rasping sounds, that the boy became alarmed and shook the palm fronds to keep them away. I could not persuade him to remain still and see what the Maiceros would do, even though he was in no peril from birds so small.

I passed the morning of July 14 watching the parents attend their three nestlings. At 6:13 the female, who had brooded through the night, left the nest and called with sharp *chip*'s from a neighboring dead tree. Two minutes later, her mate joined her there in the dim light of dawn. He flew up alone from the south, and I wondered whether he could have arrived so early from the distant roost where, apparently, many Oriole-Blackbirds slept. Soon both parents flew northward together, and a quarter of an hour later one of them brought the day's first meal to the nestlings. In the first five hours of the day, they were fed thirty-five times by both parents, who sometimes arrived together but more often came alone. After delivering a meal, one would frequently wait in a neighboring tree until its partner had arrived and fed the nestlings, after which they flew off together. From the first, the food was carried visibly in the parents' bills rather than regurgitated. Most of the articles brought to the nestlings were unrecognizable; but I distinguished small frogs on three occasions, caterpillars, earthworms, two spiders, and a black cricket.

Fairly large items were brought singly, but several caterpillars or worms were often carried together. The parents removed the nestlings' droppings in their bills, taking them just far enough to fall free of the palm tree's crown when they were released.

Much of the nestlings' food, like that of their parents, was found on the ground, where the adults walked or ran with alternately advancing feet and hopped over obstructions. Often they vanished beneath the lush pasture grass, wet with the frequent rains of this season, to emerge with their golden plumage wet and bedraggled. Their diet consisted largely of insects, caterpillars, worms, small frogs, and such other diminutive creatures as they found on the ground or amid the herbage. They varied their diet with fruits plucked from trees. In June, when a tall *Cordia*, abundant in this region, ripened multitudes of small, yellow, sweetish-astringent berries, the Oriole-Blackbirds ate many of them. Perching beside a cluster of these fruits, the blackbirds gathered them one by one, pressed the pulp from the skin, and dropped the latter. After feasting for a few minutes in the tree, they flew down to forage again on the ground. The name Maicero, or Tordo Maicero, given this bird in Venezuela, refers to its habit of eating maize. Since the corn had not yet formed ears while I was among them, I did not observe this.

I did not learn where the local Oriole-Blackbirds slept, but Paul Schwartz, who contributed so much to Venezuelan ornithology, told me that he had found hundreds roosting together in swampy woods. The nest in the palm was situated on the flyway between the locality somewhere to the south where the blackbirds slept and the foraging or breeding areas of a number of them farther north. Morning and evening, they passed over the pasture, and sometimes, principally in the mornings, they interrupted their journey to rest briefly in the trees near the palm where the nest was hidden. Usually the parents seemed not to oppose these visits, but sometimes they were mildly antagonistic toward the travelers. Early one morning in mid July, when ten blackbirds perched briefly near the nest and then continued northward, the resident pair did not protest. Half an hour later, two more Oriole-Blackbirds, traveling in the same direction, alighted in a neighboring dead tree just as the mother was about to feed her nestlings. She at once flew to the intruders from the palm, while the male did so from a nearby tree. The parents "sang" with fluffed-out plumage and seemed somewhat annoyed, but they did not try to chase the trespassers, who after a minute or two resumed their journey. This exceedingly mild demonstration was the strongest manifestation of territorial exclusiveness that I witnessed. While building, incubating, and feeding nestlings, the parents often flew far away, as though they recognized no territorial boundaries.

By July 18, when the nestlings were at most two weeks old, they had vanished. Even if they had left the nest spontaneously, they could hardly have flown far. From the hilltop where the nest was situated, I looked long and eagerly over the wide expanse of surrounding pasture for some sign of them or their parents, but in vain. Apparently, the young Oriole-Blackbirds had fallen victims to the kestrels that continued to frequent the vicinity, to a pair of Yellow-headed Caracaras nesting nearby, to one of the larger hawks that from time to time appeared at this open site, or to some other of the many birds, mammals, and reptiles that pillage nests.

Among the Oriole-Blackbirds' neighbors were several other members of the same family so well represented in South

America, and I found a number of their nests to compare with that of the blackbird. I watched a Yellow Oriole build at the leafy tip of a thin, descending twig of a mango tree. Starting with an open loop of fibrous materials, she wove painstakingly downward until this became a sleeve about 1 foot (30 centimeters) long, then closed the bottom of her pensile pouch. The quite different nest of the Orange-crowned Oriole was a pocket or shallow pouch attached to the leaflets of a large, dead frond of a yagua palm which had broken and hung vertically, swinging in the breeze. This nest, on the inner side of the dangling frond toward the palm's trunk, was so well screened by the long, slender brown pinnae clustering around it that I would not have found it if the parents, both bringing caterpillars to nestlings that I could not see, had not directed my attention to it. An anomaly in a genus of skillful weavers, the melodious Troupial, Venezuela's national bird, built no nest at all but invaded the bulky mass of interlaced twigs, containing several chambers, which Rufous-fronted Thornbirds hung conspicuously from a tree. Evicting the brown, wren-sized builders, the female Troupial lined the lowest chamber and laid three eggs, as I have told elsewhere (1969b, 1977). In the same neighborhood, female Crested Oropendolas, blackish birds as big as crows, had woven seven swinging, pear-shaped pouches, each 4 or 5 feet (1 or 1.5 meters) long, in which they incubated their eggs and fed their nestlings with no help from the males, who, bowing deeply, voiced a far-carrying, liquid *whoo-oo*.

Although the Oriole-Blackbird so closely resembled some of the true orioles in plumage, it was otherwise quite different from all these neighbors of the same family. In habits and nidification, it most reminded me, of all the birds I have watched, of the Melodious Blackbird, far away in northern Central America. These blackbirds are monogamous, hunt over the ground as well as in trees, are fond of maize, and roost gregariously. With her mate's help, the female builds in a tree an open cup well plastered inside with mud or cow dung, where she lays and incubates alone three light blue eggs spotted with black. The male helps feed the nestlings (Skutch 1954). But the all-black Melodious Blackbird has a golden voice, whereas the golden Oriole-Blackbird emits screechy notes. Although certainly closer in habits, voice, and nidification to blackbirds than to true orioles of the genus *Icterus*, the Oriole-Blackbird is a unique bird, the only species in its genus. It contributes pleasantly to the diversity of a family of ninety-four species that contains birds as different as Bobolinks, who hide their nests amid northern meadow grasses, and oropendolas, who hang theirs conspicuously in high tropical treetops, parasitic cowbirds, who drop their eggs into nests of other birds, and orioles, who carefully attend their young in exquisitely woven pouches.

13. *A Difficult Choice*

The attractions that draw us to the living world of animals and plants are subtle and difficult to analyze. If we have had the good fortune in early childhood to come into close contact with living things of varied kinds, we are often bound to them by a pervasive sympathy that joins life to life. We are already strongly attached to the natural world before we are prepared to ask just what are the ties that bind us to it. When we begin to question, we may discover that we are attracted to living things by their beauty, by the fascination of learning their ways, by the tranquility we often find in close association with them. Beyond all this, we are bound to them simply by sharing that mysterious thing called life. Many of us come to love the living world with a deep, intense, unselfish love.

That which we love we seek to join to ourselves by multiple ties. The more numerous and the more massive these bonds, the more we feel that the loved object is truly ours. The chief of these ties are sympathy, harmonious association, and knowledge. Sympathy, which is spontaneous and instinctive, is of these three the most difficult to cultivate by purposeful endeavor. Like inspiration, it may come to us unsought, or we may vainly strive to kindle it. Harmonious association and knowledge are, on the contrary, capable of deliberate cultivation. The former is what we mean by goodness, for we commonly call "good" the person or thing that enters into harmonious relations with ourselves, satisfying some vital need or helping us attain some cherished goal. "Bad" we apply to that which injures, thwarts, or conflicts with ourselves. The whole purpose of ethics is to make our striving for goodness of this sort steady, rational, and impersonal rather than capricious and self-centered. Insofar as we love the living world, we endeavor to associate harmoniously with it, to lead our lives and fulfill our aspirations with the least possible interference with other living things—in short, to cultivate an ethical relationship with nature. The more we succeed in doing this, the more we feel at one with it.

Knowledge also binds us to its object. Much of our knowledge of nature has been acquired for purely utilitarian ends, for the purpose of satisfying our basic needs more adequately and easily and even of multiplying needless luxuries. But over and above this economic motive for learning about nature, there is a more spiritual motive which doubtless always existed and grows stronger every year. It is not only, as Aristotle said, that we love knowledge for its own sake. We also love knowledge for the feeling of intimacy with the thing about which we know, for the bond it creates between this object and ourselves. We may, for example, be attracted to a bird by its beauty or its blithesome song. But it flashes across our delighted vision, its mellow notes fade away, and we have lost all immediate contact with it. If we can discover how it builds its nest, how it rears its young, or where it passes the cold winter months, we seem to have established other ties with it, to have acquired a firmer grasp of it. Thus love leads to the desire for knowledge, and knowledge binds more firmly the bonds that love creates.

In these ways, love of living things leads us to cultivate two of the most precious and honored of human ideals, moral goodness and knowledge. We wish to live in harmony with them and we wish to know about them, to understand their ways of living. In many other fields, men cultivate simultaneously the ideals of knowledge and goodness without finding any conflict between them; but those who aspire to both of these goals with reference to the living world soon come face to face with a baffling dilemma.

The sensitive student of life is involved in a tragic contradiction that rarely troubles those who pursue other branches of knowledge. The astronomer learns about the heavenly bodies without the least interference with them; he could not break them up for analysis if he would. The physicist investigates the behavior of matter in masses or minute particles without feeling that she is destroying that which she is powerless to create. The chemist knows no twinge of conscience as he dissolves salts or minerals to learn their composition. The geologist who tears apart the strata of the Earth to uncover the fossils that lurk within may destroy unique structures, but at most she barely scratches the planet's broad face. But the biologist who kills living things to discover certain facts about them, or even upsets their normal way of life to learn other facts, sacrifices the ideal of perfect goodness for the sake of knowledge. If he is morally sensitive as well as inquisitive, he cannot fail to feel the conflicts involved in his researches.

I wish it were possible to give every boy and girl who intends to become a zoologist or medical researcher or to enter some allied profession full realization of what lies before them. Perhaps love of animals, or spontaneous sympathy with them, has engendered the wish to know more about them, and they have taken the most obvious mode of satisfying this desire. Doubtless, to cut open a living earthworm or even to dissect an anesthetized frog does not seem a heartless or wicked thing to do and does not clash with their affection for dogs or horses, for furry creature or for birds, which led them into this study. If they go on to more advanced courses and are set to dissect warm-blooded animals for which they feel greater sympathy, their aversion to the occupation may become more intense. By easy steps we are led to do with hardly a qualm that which we at first never imagined ourselves capable of doing. Almost before he is aware of the changes that have occurred within him, the boy who delighted in living birds and staunchly opposed the destruction of their nests has become a professional ornithologist, perhaps taking thousands of feathered lives in the name of science. Or the youth who hesitated to cut into an anesthetized frog is performing, on living dogs and monkeys, experiments that make us shudder. And these people to whom the killing and mutilation of living creatures are routine occupations are no longer free to consider with calm detachment the full implications of their activities. Their daily bread, the welfare of their families, may depend upon their continuance. Moral judgment may be so strained by economic needs that it is no longer dependable.

It is a pity that the great teachers to whom large sections of humanity look for guidance failed to consider this conflict between goodness and knowledge which had hardly become a problem in their day. It is likewise regrettable that more recent philosophers and moralists, with the notable exception of Regan (1983), who have written at such great length on a wide range of ethical questions have not given this matter the attention it demands. But it is not hard to imagine how some of the revered prophets and sages of old would have treated the question. I have little doubt how Mahavira, the lawgiver of the Jains, and Gautama the Buddha would have answered us. When they forbade their followers to take life, they were not thinking, as in the Mosaic code, merely of men of their own nation, but of all animate creatures. It is highly improbable that they would have made exceptions to their rule when it was pointed out to them that it is sometimes necessary to take life in order to learn how animals are constructed, how they function, and how they should be classified. They would have told us roundly that the first thing it behooves us to know is ourselves, which we begin to do when we free our minds of all blinding passions and gaze steadily upon our origin and our destiny. They might have gone on to point out that, once we understand ourselves, we shall also know as much as is necessary about other creatures, for all living beings are fundamentally the same. Lao-tzu, the Taoist sage who said that the best of men are like water, that benefits all things and does not strive with them, would, I fancy, have returned much the same answer.

I find it more difficult to imagine how Jesus would have treated the problem. Apparently, he believed that the world order, as known to us, was fast approaching its end, and it would have been consistent with this view to hold that knowledge of natural processes is no more necessary for gaining the kingdom of heaven than the possession of worldly goods. Unfortunately, the whole subsequent attitude of the Western world to this matter was determined not so much by Jesus as by St. Paul, an extremely able propagandist of rather narrow sympathies. Since he questioned whether God could care about an ox (I Corinthians 9:9–10), he most probably would have maintained that men need have no compunction about gathering whatever knowledge they find useful or agreeable, without pausing to consider how their researches might hurt nonhuman creatures. For the whole "brute" creation of the Western world, the Pauline attitude has had tragic consequences.

Since we search in vain through the pronouncements of mankind's most revered teachers for adequate consideration of the conflict between the ideals of goodness and knowledge that confronts those who associate most intimately with the living world, I suppose that each of us must ponder the problem for himself and take his own stand. After years of questioning, I have taken mine. But my present purpose is not to offer a definitive answer to the dilemma, to propound or defend any special view, so much as to clarify the alternatives. Too long have those who call themselves friends of nature refused to look squarely at the implications of their position; too long have we slurred over the contradictions involved in it or hastily accepted conventional compromises, which, when examined, are found to rest upon the flimsiest of foundations.

I should be sorry to create the impression that I see an irreconcilable opposition between the goal of perfect goodness and that of complete knowledge—between the ideals of religion and of science when carried to their ultimate logical conclusions. To find these highest and noblest human aspirations radically incompatible might make us lose faith in the unity and soundness of our nature. On the contrary, I hold that the more adequate our knowledge, the more completely we can realize our ideal of goodness, and the greater our goodness, the more perfect our understanding becomes. It is not knowledge itself but the means that beings with our peculiar limitations in sensory and mental equipment are often driven to employ in the pursuit of knowledge, especially that of the living world, that so often cause biologists and other scientists to violate the ideal of goodness. It is conceivable that beings with more penetrating minds and an ampler endowment of senses than we have could learn all that we aspire to know without harming any creature.

Too often we take the shortest and easiest way when a more painstaking method would yield not only the information that we desire but bring us fuller understanding in the end, all without injuring the creatures that we investigate. While studying the nesting habits of certain species of birds in which the sexes cannot be distinguished by appearance or voice, I have sometimes wished to learn which member of a pair was the male and which the female. Two ways were open to me—to shoot one of the birds and perform an autopsy, or to see which of the two laid an egg. The first method might have yielded the desired information in a few minutes; the second sometimes required many hours of careful watching, but by it I learned things of great interest that would have escaped me had I taken the easier way. In investigating the resistance of animals to climatic extremes, we can, if we have the necessary expensive equipment, confine them in freezing chambers or heated compartments until they succumb, or we can observe how different climates and extremes of weather affect them in their free state. The first method is quicker and easier, but the second may yield the fuller knowledge.

We can base our anatomy, and the classification that rests upon it, on the study of animals killed for this special purpose, or limit our researches to those that die by means beyond our control, as the human anatomist must. Again, the first is quicker, but the second is more satisfactory to an exacting conscience; and perhaps there is no great urgency in this matter. Facts of anatomy and physiology, which today it seems impossible to learn without killing or maiming animals, may tomorrow, with improved apparatus and methods, be learned from living creatures that suffer no harm. For testing the effects of drugs, eggs can often be substituted for laboratory animals. Some researches for which biologists are willing to torture or to sacrifice large numbers of animals are directed to questions of doubtful importance.

Since, once we pass beyond the narrow sphere of human society, our conventional religions and philosophies fail to provide guidance, each of us must decide for himself whether goodness or knowledge is to have precedence, whether it is more important, or more satisfying, to cultivate harmonious relations with living things or to know about them, even at the expense of such harmony. In reaching a decision on this pressing problem, we shall, no doubt, be influenced by the most diverse considerations, but two seem to merit particular attention. The first is that of completeness, of the possibility of attaining the goal we set for ourselves. We must admit at the outset that neither perfect goodness, in the sense of harmony in every aspect of our lives, nor complete knowledge is, for beings such as ourselves, an attainable ideal but at best a limit toward which we can advance by an endless progression. Since we cannot eat without destroying some living thing, can hardly take a step in the open fields without crushing some minute creature, it is obvious that, so long as we live and move, we cannot realize that ideal of goodness which consists in harming no organized being.

Although in past ages the savant could make all recorded knowledge his province, in modern times the growth of the sciences has been so rapid that the most capacious intellect can hardly encompass a single one of them; yet all mankind's actual knowledge accounts for but a small fraction of what might be known. In comparison with the number of facts we would need to acquire in order to possess complete knowledge, even of our solar system, the number of our daily contacts with other beings is small. By taking thought to make these contacts more harmonious and mutually profitable, perhaps also by reducing their number by simplifying our lives, we can draw ever nearer to our ideal of goodness. Although neither complete knowledge nor perfect goodness is attainable by us, we can come much closer to the latter than to the former, and it seems wise to strive toward the goal which we can most nearly reach.

The second consideration is that of intimacy, of whether knowledge or goodness is more central to ourselves, less likely to be lost after it has been won. At this point, it may be profitable to recall that we *have* knowledge but we *are* good. In becoming good, we improve ourselves to the great advantage of the beings around us; in learning facts, we amass possessions that may be lost by forgetting. Our gains seem more secure when we refine and ennoble our own nature by living in concord with the things around us than when we merely learn about them. This harmonious association binds us to the living world more securely than knowledge can and more completely satisfies that love of living things which led us to consider this perplexing problem.

14. *The Scaly-breasted Hummingbird*

December 25, 1935, dawned with a brisk, chilling breeze driving down the narrow valley of the Río Buena Vista from the high summits of Costa Rica's Cordillera de Talamanca at its head. I decided to celebrate Christmas Day by resting from the plant collecting that then engaged me and looking for birds. The early morning was brilliantly clear as I walked down the rocky, grass-covered cart road toward San Isidro. The bushy pastures and thickets along the way were bright with a profusion of flowering shrubs, among which yellow and white composites were prominent. Amid this dense cover the birds lurked in silence, as they usually do at the beginning of the dry season. I saw scarcely any until I came upon a singing assembly of rather large and exceptionally plain hummingbirds new to me.

Their upper plumage was metallic bronzy green. Their tails were more bluish green, with the two or three outer feathers on each side tipped with dull white, the outermost most broadly. Behind each eye was a small whitish spot.

On the throat and breast the feathers, dull green with grayish buffy margins, resembled scales at close range. The abdomen was pale brownish buff. The fairly long straight bill was black, with pale red at the base of the lower mandible. Later, I learned that the females of this species closely resemble the males. Since I lacked a guide that described this bird, I called it the Christmas Hummingbird, under which name I continued to collect notes about it until, a decade later, I visited the United States. There I learned that it is known as the Scaly-breasted Hummingbird, a species which ranges from Guatemala to Colombia and from sea level up to about 4,000 feet (1,220 meters) in Costa Rica. Avoiding closed woodland, it prefers areas with moderately tall scattered trees, such as shady pastures, lightly shaded coffee plantations, roadsides, dooryards, and open second-growth woods.

The singing assembly consisted of at least four of these modestly attired hummingbirds, who perched from 20 to 40

Scaly-breasted Hummingbird

feet (6 to 12 meters) up on exposed dead twigs of trees standing above the tangled, weedy growth that pressed close to the roadway on both sides. Three of the participants occupied the points of a roughly equilateral triangle with sides about 100 feet (30 meters) long, while the fourth hummingbird was stationed about the same distance beyond one of the apices. Three were in trees beside the road, the fourth off in a bush-choked field.

As these Scaly-breasted Hummingbirds perched on naked twigs, they repeated at irregular intervals a little phrase that was certainly not the least melodious of the hummingbirds' songs that I have heard. I noticed some variation among the verses of the four songsters and even between successive songs of the same individual. That which I most frequently heard sounded like *cheee twe twe twe twe*—trill—*chup chup*. The trill, which first caught my ear, so closely resembled the little trill of the Black-fronted, or Common, Tody-Flycatcher that I looked for this diminutive yellow-breasted bird and was surprised to find that a hummingbird was the author of the clear notes.

In addition to the foregoing verse and its numerous variations, these hummingbirds delivered a very different kind of song. It was a sort of medley composed of the smallest possible notes, too slight even to be called squeaky. Indeed, much of this second song was too weak to be audible to me at a distance of 20 or 25 feet (6 or 7.5 meters), if not too high in pitch to be perceived by the human ear at any distance. Yet after this tenuous medley ran off into silence, I could see by the hummingbird's outswollen, vibrating throat that he was still singing. Such whisper-songs are not rare among hummingbirds. The performance of these Scaly-breasted Hummingbirds, especially their scarcely audible verses, reminded me of the more richly varied song of a much smaller species, the Wine-throated Hummingbird of the Guatemalan highlands.

As the drought increased in January, this singing assembly of Scaly-breasted Hummingbirds dispersed. At the end of June, when the rainy season was in full sway, I noticed that it was active again. In May of the following year, 1937, I again found the hummingbirds performing in the same place, so this assembly was maintained during at least three wet seasons.

In later years, I have found many more Scaly-breasts performing in the same manner, usually on exposed twigs of small or middle-sized trees that stood above tangled lower growth, but often also in our garden or on shade trees of coffee plantations, well above the tops of the coffee shrubs but rarely more than 40 feet (12 meters) above the ground. I have not found more than four individuals in an assembly; sometimes only two perform within hearing of each other, and occasionally one proclaims his presence alone. The same stations are occupied year after year—if all goes well, for a decade or more. Hummingbirds' songs appear to be learned rather than innate, and, as in other species, those of Scaly-breasts vary from assembly to assembly. Many lack the clear trill that was so noticeable at the first assembly. A frequent version might be paraphrased *see seea chweee, see sea chweee, see sea chip-chip-chip-chip-chip*. The dry, rapidly repeated *chip* apparently represents the trill of other individuals. The *chweee* is a low, full note, exceptionally strong and mellow for a hummingbird, clearly audible at a distance of 200 feet (61 meters). Between songs the hummingbirds sometimes fan out their tail feathers, revealing the prominent whitish tips of the outer ones, and wag their spread tails vigorously while they shake their relaxed wings.

Scaly-breasts perform in their assemblies throughout the day, from early morning until late afternoon. In this they agree with many other species but differ conspicuously from another that shares their habitat, the Rufous-tailed Hummingbird, which sings methodically at dawn but rarely after sunrise.

In the Valley of El General, Scaly-breasts are often silent during the early part of the year, when the weather becomes increasingly dry and often, by March, the vegetation languishes in the desiccated soil. Sometimes they resume singing in April, when showers have refreshed the plants; and usually they are in full song by early May, when rains have become frequent and the vegetation is again lush. Sometimes the Scaly-breasts continue to perform through the long wet season until the year's end, but occasionally I have failed to hear them after September. The period of song by the males coincides closely with the nesting season of the females; and it seems clear that, as in numerous other hummingbirds, they perform to advertise their presence to the other sex rather than to proclaim possession of a feeding territory. The songsters frequently fly beyond view to forage. But, as in other hummingbirds with singing assemblies, I have not succeeded in following courtship to its consummation. In this species, it is impossible to tell whether another individual who approaches a songster on his perch and is rapidly pursued well beyond sight is a female or an intruding male.

Like other hummingbirds without highly specialized bills, the Scaly-breast sucks nectar from a variety of flowers, large and small, and it likewise catches minute insects in the air. In our dooryard are four trees of the Red Poró, which toward the end of the rainy season drop their trifoliolate leaves and display masses of scarlet flowers on nearly naked branches. Each larger tree is usually claimed by a Long-billed Starthroat, whose bill serves well to extract nectar from the base of the long, tightly folded, sword-shaped standard, the only part of the corolla of this highly modified papilionaceous flower that is well developed and exposed to view. These big hummingbirds drive away smaller species which can only surreptitiously visit the scarlet flowers. Those with bills too short to reach the nectar in the usual way procure it by piercing the thick, tubular calyx that forms a collar around the base of the standard. This is regularly done by the sharp-billed Purple-crowned Fairy and often, too, by the Scaly-breasted Hummingbird, while it either hovers on the wing or clings to the end of the standard. At other times it reaches the nectar from in front of the flower, spreading apart the lower edges of the folded standard to push its bill far inward.

In late November, two Scaly-breasts contended for possession of a poró tree not defended by a Starthroat. Sometimes they seized one another and fell together to the ground, where after a few seconds they separated to rise into the tree again. Here they might perch only a few inches apart, resting with tails partly spread, for as long as ten minutes, until one dashed at its adversary. Then the two would dart around, uttering low squeaks and clashing at intervals. They pulled feathers from each other; one had a tuft of its opponent's down clinging to its bill for many minutes. This bird had disheveled plumage, but the other who had lost the feather was even more bedraggled. The hummingbird whose plumage was less tattered was apparently a male, for while he perched with his closed bill pointed toward his adversary, his throat swelled out and vibrated; he seemed to be singing in an undertone, as male hummingbirds frequently do. After a few days, the other vanished.

Except at their singing assemblies, when males who compete for females cooperate to attract them to a traditional site, hummingbirds are solitary and aggressive, frequently chasing each other and driving from flowers nectar drinkers of their own and other kinds, including larger insects. Such aggressiveness is widespread among nectarivorous birds. In New Guinea, avian assemblages at flowering trees were "veritable riots of interindividual aggression" (Beehler

1980). The Singing Honeyeater of Australia is extremely pugnacious, fiercely attacking birds of all kinds, especially those smaller and weaker than itself. Moreover, it eats the eggs of smaller birds (Goodwin 1978). Pairs of Amakihi, a Hawaiian honeycreeper, subdivide their territories while foraging, and the female actively defends her flowers against her mate (Kamil and van Riper III 1982). Even wood warblers, usually insectivorous, become more aggressive when they drink nectar, as has been recorded of Cape May and Palm warblers in the Bahama Islands and Tennessee Warblers in Costa Rica (Emlen 1973; Tramer and Kemp 1980).

It is instructive to compare the nectarivorous hummingbirds with the largely frugivorous tanagers, about equally numerous in species and individuals in tropical America and equally brilliant. Tanagers live throughout the year in small flocks or inseparable pairs and are among the mildest of birds. At a tree or shrub laden with berries, one may watch many tanagers, honeycreepers, finches, flycatchers, manakins, cotingas, and others all peaceably eating together, with rarely a hostile gesture. Among honeycreepers, the most aggressive is the Green, which takes much nectar in addition to fruits.

Why this difference in temperament between nectarivorous and frugivorous birds? It may be because a fruiteater can see immediately, by its color, whether a berry or other fruit is ripe and does not waste time sampling green ones. Without inserting its bill into a corolla, a hummingbird probably cannot tell whether a flower contains nectar or has recently been drained by another. The energy derived from the nectar of many a small flower may exceed only slightly that spent by the hummingbird in flying to and hovering before it, so that the bird cannot afford to probe many flowers that yield nothing. To avoid this, it tries to exclude other visitors from the flowers that it claims. If it remembers the flowers that it has recently visited and does not return to them before they have had time to replenish their nectar, this exclusiveness may benefit it greatly.

While watching a manakin's nest from a blind set amid tall second-growth woods not far from a clearing, I saw a Scaly-breasted Hummingbird bathe on a soft cushion of green moss attached to a slender stem. The hummingbird clung to the water-soaked cushion, pressed in its breast, rubbed its face against the moss, then perched nearby and shook its relaxed feathers. A little later, a Rufous-tailed Hummingbird bathed there in much the same fashion. This appeared to be a recognized bathing place of the hummingbirds, for three weeks earlier I had seen a Rufous-tail wet its feathers on this cushion of moss. One frequently sees hummingbirds as well as other feathered creatures perform their ablutions amid foliage laden with dew or raindrops, but the use of a sponge of moss seems to be less usual.

In the Valley of El General, Scaly-breasted Hummingbirds start to build in May after the rainy season has become well established, but of the twenty-seven nests that I have found, only one received eggs so early. In June, when eight nests held eggs, the breeding season reaches its peak. July, August, and September each had four nests with eggs, and I found one or two in each of the three remaining months of the year. In January, when flowers are still abundant in the sunnier days at the beginning of the dry season, late nesters continue to feed young from eggs laid in December, but I have found only one set of newly laid eggs. From this exceptionally late nest, the nestlings flew on February 11, when in many years flowering plants become much scarcer in the desiccated soil of the clearings.

Scaly-breasts build their nests from 6 to 25 feet (1.8 to 7.6 meters) up, mostly between 10 and 20 feet (3 and 6 meters), in small trees growing in shady pastures, plantations, and dooryards. One nest was beside a stream, and two were above water. Of the latter, one was 8 feet (2.4 meters) above the rocky channel of a mountain torrent, 25 feet (7.6 meters) from the shore. The nest is often placed on an upright or ascending branch between several diverging twigs that hold it firmly, but sometimes it rests on a thicker horizontal branch, with perhaps one or two upright shoots to provide lateral support. At times the nest is partly screened by leaves, but often it is in a very exposed situation, even on a leafless dead branch. The branch that supports the nest may be covered with moss or liverworts with which the structure blends, so that even in the absence of concealing foliage it is not conspicuous. One nest was built on the horizontal part of the dangling loop of a slender woody vine about 25 feet (7.6 meters) above the mouth of a creek that flowed into a wide mountain stream. Well covered with gray lichens and green moss, the nest appeared to be an excrescence of the vine.

As is usual among hummingbirds, the female builds with no help from her mate. She works chiefly soon after sunrise, gathering her materials from trees rather than the ground. I watched one Scaly-breast loosen lichens from a trunk by inserting her bill beneath their edges and prying them up while hovering on the wing. Twelve visits to the nest in a quarter of an hour was the most rapid building that I have seen. When finished, her nest is a beautiful structure. Before the growing nestlings spread it outward, it is sometimes almost hemispheric, with a strongly incurved rim. Other newly finished nests are broader and shallower. Composed largely of soft, downy materials, the nest is more or less completely covered on the outside with green liverworts, green mosses, and light gray or blue-green foliaceous lichens. Usually one of these types of covering predominates, with an admixture of the others.

When mosses are used to cover the nest, they may hang in streamers beneath it. Often whitish foliaceous lichens line the inside of the nest as well as cover its exterior. At times fine fungal rhizomorphs—"vegetable horsehair"—are mixed with the down that composes the body of the nest. One nest measured 2 inches in overall diameter by 2.25 inches in height. The inside diameter was 1.5 inches (5 by 5.7 by 3.8 centimeters).

Soon after she finishes her nest, the female lays her first egg, and two days later, around sunrise, she deposits the second. Nearly always she lays two; an occasional nest with only one may have lost another. The unmarked white eggs are long and narrow, as is typical of hummingbirds. Sometimes the Scaly-breast sleeps over the first egg for one or two nights before she lays the second, which will make a substantial difference in the times they hatch. When sleeping, whether on a nest or perching on a twig, hummingbirds hold their heads forward and inclined slightly upward instead of turning them back among the feathers of their shoulders, as many birds do.

In mid September, I passed a whole forenoon and all the afternoon of the following day watching a nest in which the second egg had been laid eleven days earlier. In the morning, brief showers fell before the sun broke through the clouds around eight o'clock. The afternoon of the following day was at first clear, but after two o'clock rain and drizzle fell most of the time until nightfall. This Scaly-breast's active day began at 5:47 A.M. and ended at 5:42 P.M. In twelve hours, she took thirty-five sessions on the eggs, ranging from less than 1 to 103 minutes and averaging 14.3 minutes. An equal number of recesses ranged from less than 1 to 23 minutes and averaged 6 minutes. She covered her eggs for 70.4 percent of the day. She sat far more constantly in the afternoon than in the forenoon. Before midday she left her nest thirty-one times, but after midday she did so only four times (not counting two absences of two or three minutes each, when she left as domestic animals passed beneath her). In the forenoon her longest continuous session was 30 minutes, whereas in the afternoon she sat once for 103 minutes and twice for 69 minutes. Her longest absence in the forenoon was 23 minutes; in the afternoon, 20 minutes. Her interval of greatest restlessness was from seven to eight o'clock, during which she left her nest and returned to it seventeen times. Her longest session during this hour lasted 11 minutes, and many of her sessions and absences continued for less than 1 minute.

This hour of greatest restlessness was when the Scaly-breast most actively added material to her nest. On at least thirteen of her returns she brought either lichens or cobwebs; probably each time she came back from a brief excursion she brought something for her nest, but at times it was too small to be distinguished in her bill. The lichens were always larger or smaller fragments of a whitish, foliaceous species. Sometimes the hummingbird tried to fasten them to the outside of her mossy nest, which was already so thickly encrusted with them that no more would stick there. Accordingly, in the end she always placed them inside the cup with her eggs or else on the rim. When she brought cobwebs, she settled in the nest and then bent her head down to run her long bill over the outer surface, spreading the silk over the lichens that covered it. After this operation, she usually bounced up and down in the nest, a movement that suggested that she was kneading the nest's lining with her feet. After the middle of the morning, she incubated more constantly and brought much less material to her nest. Her last contribution of the day, a tiny piece of lichen, was brought when she returned to resume sitting at 11:12 A.M.

The Scaly-breast's different behavior in the morning and afternoon might be attributed to the circumstance that at this season long, hard afternoon rains were frequent, whereas the mornings were often sunny. However, many hummingbirds take time from incubation to add to their nests in the morning rather than the afternoon, even at seasons when rain is infrequent.

To leave her eggs, this Scaly-breast often flew backward until clear of the nest, then reversed her direction and darted away. Rarely she uttered a few sharp notes while sitting. All day I neither saw nor heard a second Scaly-breasted Hummingbird that might have been her mate.

Another female Scaly-breast incubated more constantly. In twelve hours of observation she took only thirteen sessions that ranged from 4 to 99 minutes and averaged 38.5 minutes. Her absences varied from 1 to 34 minutes and averaged 9.3 minutes. She covered her eggs for 80.5 percent of her active day. Hummingbirds in general incubate more constantly than many larger birds, probably because their diet of nectar supplemented by tiny insects is so sustaining.

At one nest the incubation period, counting from the laying of the second egg to the hatching of the last nestling, was no more than seventeen and a half days. More often it was between eighteen and a half and nineteen days—about 50 percent longer than the incubation period of the much larger eggs of many thrushes, tanagers, sparrows, and other songbirds. In a rainy September, a Scaly-breast's egg took nineteen or twenty days to hatch, and its companion failed to hatch.

At one nest, both eggs hatched within an interval of six hours, and at another they hatched within ten hours. The female hummingbird does not, like most passerine parents, promptly remove the empty shells. Usually they stay in the nest for hours. Sometimes they finally disappear, apparently having been thrown out by the parent, possibly unintentionally, by being caught among the feathers of her abdomen when she flew away. Frequently the shells remain until they are flattened out and broken into fragments. Probably the papery white shells are often overlooked by the parent because they do not contrast with the whitish lichens that frequently line her nest.

As has often been remarked of other kinds of hummingbirds, the nestlings break out of the shell in such an undeveloped state that, unless closely examined, they resemble insect grubs. Newly hatched Scaly-breasts have very short bills, tightly closed eyes, rudimentary wings, black skins, and, along the center of the back, two parallel rows of pale gray down-feathers, about eight in each row.

In the rainy season of 1962, a Scaly-breast nested in the top of a Calabash tree in front of our house. Although her nest was in a most exposed situation, it blended so well with the moss and liverworts that thickly covered the supporting limb that I failed to notice it until several days after the eggs had been laid. This nest was unusually favorable for study, not only because of its location but also because its attendant had a conspicuous swelling near the base of her lower mandible which served to identify her. As has long been known, male hummingbirds nearly always neglect their progeny; but after Moore (1947) and Schäfer (1954) reported that in South America the male Sparkling Violet-ear helps the female to attend the nest, it became desirable to investigate this matter in as many species as possible. In the many hummingbirds of which the sexes are difficult to distinguish on the wing, an occasional visit to the nest by a male might be overlooked. Likewise a female helper, such as Wagner (1959) found at a White-eared Hummingbird's nest in Mexico, might escape detection if her visits did not coincide with those of the mother, who might chase this unwanted assistant away. But at this nest attended by a female with a wart on her bill, I could hardly fail to recognize a second attendant of either sex if one arrived.

On August 14, when the two nestlings in the Calabash tree were five and six days old, I passed the morning watching their nest from the porch. In the first six hours of the day, the parent with the wart fed the nestlings fourteen times, and no other hummingbird came near them. Because their nest was well above me, I could see little of them; but it was evident that on each parental visit they were fed in several installments, or separate acts of regurgitation, between which the mother's bill was removed from the nestlings' mouths. The number of installments which the parent delivered on a single visit varied from about three to eleven, and the total time devoted to feeding ranged from about twenty seconds to slightly over one minute, as measured by a stopwatch.

In the six hours, the parent brooded the nestlings twenty times, for intervals ranging from less than 1 to 26 minutes, totaling 189 minutes and averaging 9.5 minutes. Her twenty-one absences from the nest, ranging from 1 to 18 minutes, totaled 167 minutes and averaged 8 minutes. The nestlings were brooded for 53 percent of the morning. The longest sessions of brooding were not in the cool early morning but after ten o'clock, when sunshine was falling brightly on the fully exposed nest and the naked nestlings needed protection from it. At this time the parent brooded more often than she fed. Often she opened her bill to pant in the hot sunshine, as her nestlings did when she left them exposed to it.

When these nestlings were only nine and ten days old, with sprouting pinfeathers that failed by far to cover their skins, their mother no longer brooded them by night. Such early cessation of nocturnal brooding is not unusual in Scaly-breasted Hummingbirds. At another nest, the two young slept alone when they were only eight and nine days old and again on the following night. On the next night, however, their mother brooded them, although no rain was falling as the day ended. This was the last time that I found her on the nest at night; nocturnal brooding definitely ceased when these nestlings were eleven and twelve days old. At another nest, the parent stopped brooding when the nestlings were nine days old; at still another, when they were eleven days old.

In other hummingbirds that nest in mild climates, brooding stops at a similarly early age. Anna's Hummingbirds in California were not brooded either by day or by night after they were thirteen days old, at which age they maintained their temperature well above that of the air at all times (Howell and Dawson 1954). But at high altitudes, where frost is frequent at the season when many hummingbirds nest, brooding may continue considerably longer. White-eared Hummingbirds, breeding in November and December at 8,500 feet (2,590 meters) in the Guatemalan highlands, covered their young by night until they were seventeen or eighteen days old and well feathered.

Surprisingly, Scaly-breasts cover their nestlings during a daytime shower long after they have ceased to brood them through the night, even after they are clothed in plumage. Yet nocturnal rain is by no means absent at the season when these hummingbirds breed, although the hardest downpours usually come in the afternoons. Probably if rain were falling hard in the evening the parent would remain on the nest through the night, but while the nestlings are still nearly naked they are liable to be exposed to a rain that starts after nightfall. A mother of two thirteen-day-old nestlings left them exposed to a shower that was falling as the day ended. However, as I learned long ago, nestling hummingbirds are unbelievably hardy (Skutch 1931, 1981).

On August 24, when the two nestlings in the Calabash tree were fifteen and sixteen days old and fairly well feathered, I again watched through the morning. In the first six hours of the day, they were fed nineteen times, only a small increase over the number of meals they had received ten days earlier. On each feeding visit the parent, standing on the nest's rim, delivered food in four to eight separate acts of regurgitation, but when as many as eight were made, each was brief. Sometimes I clearly saw that the parent regurgitated to the two nestlings alternately, but because of the nest's height I could not always see plainly. On a single visit, food was delivered over an interval ranging from about twenty-five seconds to one minute. Once the parent brought a small object, apparently an insect, in her bill and gave it to a nestling along with the regurgitated food. Hummingbirds rarely deliver visible objects.

Although the mother with the wart on her bill had long since ceased to brood in rainless weather and by night, for two minutes early in the morning she rested on the nest's eastern rim and shaded her nestlings from the rising sun. Otherwise, she made no effort to shield them from its rays, which fell upon them until late in the morning, when the nest was in the narrow shadow cast by a higher mossy branch of the Calabash tree. When exposed to the bright sunshine, the nestlings stretched up their heads and panted with open bills. The ability of these diminutive creatures to resist strong insolation was amazing. Many years ago, a feathered nestling of a Spotted-breasted Wren that I had placed in the sunshine for photography unexpectedly succumbed to a much shorter exposure than these young hummingbirds resisted morning after morning.

I also watched the nest in the Calabash tree through the afternoon of August 28, when the nestlings were nineteen and twenty days old. In the first hour, they were fed three times. Then a moderate shower fell for ten minutes, during which the nestlings rested with their bills pointing almost straight upward. After an intermission, another shower began. While it fell, the parent fed the nestlings and then tried to brood them, but they persisted in keeping their heads exposed in front of her breast, and one flapped its wings. During the nineteen minutes while rain fell rather fast, all three, mother and nestlings, held their heads tilted sharply upward, and the latter sometimes kept their mouths open, as though catching the raindrops. When the shower diminished to a drizzle, the parent left to collect food. After receiving a meal, one of the young birds rose up in the nest to flap its wings vigorously. From three to nearly five o'clock, the mother was mostly out of sight, and the nestlings received nothing. Then she returned, and in the last eighty minutes of the day she delivered six meals, sometimes feeding the nestlings alternately, sometimes in the order ABBA. Occasionally one young hummingbird received three installments and the other only one. The last meal was delivered at 6:13, in the dusk when bats were flying.

On other evenings, too, these nestlings were fed generously well after six o'clock, when the light had grown dim. At other nests I have watched Scaly-breasted Hummingbirds feed their young when the evening twilight was so far advanced that I could scarcely distinguish the parent, even through my binocular. Nearly all the diurnal birds of other kinds had by this hour gone to rest and fallen silent. This accelerated feeding in the last hour of the day, continuing until it is nearly dark, fortifies the nestlings against the long night when they remain unbrooded and unfed.

In six and a quarter hours of the afternoon, the two nestlings in the Calabash tree received twelve meals, half of them in the last eighty minutes of the day. Combining this record with that made in the forenoon of August 24, we find that two feathered Scaly-breasted Hummingbird nestlings, fifteen to twenty days old, received thirty-one meals in a day of twelve and a half hours. The parent brought food at the rate of 1.2 times per hour for each of them. I have rarely known hummingbirds of any species to feed more often than 2.5 times per hour for each nestling.

Nestling Scaly-breasts develop slowly. At the end of a week, they look much as they did when newly hatched, except that they are much larger. At the age of nine days, their pinfeathers start to sprout through the skin. When the nestlings are fifteen days old, they can open their eyes, but they keep them closed much of the time. At sixteen days they are fairly well clothed by their expanding plumage, but they stay in the nest a week or ten days longer. As the time for their departure approaches, they often exercise their wings by beating them rapidly while they cling to the nest with their feet to avoid being lifted from it.

Soon after six o'clock on the morning of September 2, one young hummingbird (A) left its nest in the Calabash tree. A little later I found the fledgling resting on a dead twig at the very top of the Calabash tree, exposed to the sky. This perch, about 2 yards (1.8 meters) above the nest, was a favorite resting place of its mother, and here she fed it. Both this young Scaly-breast and the other (B) still in the nest preened much. The latter was restless, turning its head from side to side, twitching its wings, sometimes whirring them rapidly. Then, at eight o'clock, it flew from the nest to an exposed perch 2 or 3 yards (1.8 or 2.7 meters) away. At the moment of its departure, the parent was resting about a yard from the nest, but as far as I could tell she did not urge it to leave. While one young hummingbird was still in the nest and the other outside, she had fed both of them.

When, a quarter of an hour later, fledgling B flew down into a Madera Negra tree, its mother followed it closely, as many larger birds do on their fledglings' first flight, possibly to shield them from aerial attack. Nevertheless, she did not try to move her offspring to less exposed situations. Presently fledgling A flew to another exposed perch at the very top of the tree in which it had hatched. For nearly three hours it rested there, fed at intervals by its mother. Soon after noon I lost sight of both fledglings, but at one o'clock I noticed one of them perching in an Annatto tree in view of my study window. All afternoon, for more than four hours, it rested in the same spot, receiving meals from its parent. Since these fledglings rarely uttered a note to guide their attendant to them, their immobility made it easier for her to find them for successive feedings. When young birds are noisy, their voices guide their parents to them, and it is not so important that they remain stationary.

That evening I noticed a fledgling resting on a nearly leafless branch at the top of the Calabash tree, 2 or 3 yards (1.8 or 2.7 meters) from the nest that it had left earlier in the day. Here it remained until morning, without a leaf to shelter it from the rain that began soon after nightfall. I did not again find one of these young hummingbirds roosting in such an exposed situation.

I saw only the parent with the wart on her bill take an interest in these young hummingbirds while they were in the nest and after they left it. One morning, before the young fledged, a male Scaly-breast perched on a twig of the nest tree and sang for a minute or two without approaching the nest.

Eighteen years earlier, I had watched another brood of two leave their nest in a streamside tree. Both flew from the nest quite spontaneously, while their mother was out of sight, one at 6:56 A.M., the other thirty-six minutes later. In the interval when one was in the nest and the other beyond it, both were fed. The first flights of both fledglings were short, only a yard or two, but one left the nest by flying almost straight upward, an admirable feat for a bird with untried wings. Soon both made longer flights, up to about 20 feet (6 meters). They alighted without difficulty, even after their very first flights. Both of these young hummingbirds refused food soon after departing their nest. Neither they nor any other Scaly-breasted Hummingbirds that I have watched have returned to rest or sleep in the nest after their first departure. Before the young are fledged, they quite fill the nest, which their growing bodies sometimes burst asunder and flatten out. Occasionally, however, they leave the nest sufficiently intact to be used for a second brood after some refurbishing by the parent.

Of the eleven nestling Scaly-breasts whose time of departure I know, seven severed contact with their nests before eight o'clock in the morning. Three others left later in the morning or around noon. Only one is definitely known to have abandoned its nest in the afternoon. Eighteen young Scaly-breasts left ten nests at ages ranging from twenty-two to twenty-nine days; the average length of the nestling period was twenty-four and a half days. The shortest period was that of a nestling who grew up alone at the very end of the breeding season, in January and early February, when liberal nourishment at a time of little rain and abundant flowers accelerated its development. Two nestlings reared together in a very rainy October lingered in their nest for twenty-six and twenty-seven days. A lone nestling that grew up in another wet October left at the exceptionally advanced age of twenty-nine days. This nesting in a very rainy spell was in every way retarded: the second egg was laid late rather than early in the morning; only one egg hatched after an incubation period two or three days beyond the normal; and the nestling period was prolonged about four days beyond its average length. A similar retardation of nestlings' development by inclement weather is not unusual in other species. Two Rufous-tailed Hummingbird nestlings, raised at the height of the wet season, delayed in their nest for twenty-five and twenty-seven days, although in favorable weather young Rufous-tails fly at the age of about twenty-two days. Although most Yellow-bellied Elaenias abandon their nests when seventeen or eighteen days old, two young of a late brood reared in wet weather stayed for twenty or twenty-one days. The nestling period of European Swifts varies greatly with the abundance of nourishment their parents are able to provide for them, which in turn depends on the weather (Lack 1956).

For more than a month after the fledgling Scaly-breasts left their nest in the Calabash tree, I saw little of them. Occasionally I heard the sharp, rapidly repeated, staccato note that their mother used to call them when ready to feed them. One afternoon she rested on a dead twig at the top of a large orange tree in front of the house, repeating this call for several minutes without attracting a juvenile. Finally, on October 7, I saw her feed one of her offspring in a low tree near the nest. I identified her by the wart on her lower mandible, which had been shrinking since August but was still discernible through a binocular. The young bird was almost as big as his mother, but his bill was slightly shorter. I use the masculine pronoun advisedly, for reasons that will presently appear.

On October 11, 12, and 13, this young hummingbird spent much time in view of my study window. On these days I saw his mother feed him nine times, always on the same thin, exposed twig, which I shall call the feeding perch, in the Annatto tree. Even when the young bird was resting in full view a few yards away, she did not alight beside him, as she might well have done, but went to the feeding perch and called with the usual staccato notes, whereupon he flew up beside her. When he opened his mouth, she pushed her bill into his throat and regurgitated from one to four times, with much violent shaking of the heads of both. Then she would fly away. Sometimes he also left, but at other times he stayed in the Annatto tree, resting for long intervals on the feeding perch itself. Once, when he was out of sight when she arrived, she waited on the feeding perch, repeating a single low note. After five minutes, the young Scaly-breast returned and received a generous meal. Having a definite spot for the delivery of meals doubtless helped the adult maintain contact with the young hummingbird, whose calls were weak.

At 11:20 A.M. on October 11, I found the juvenile in the Annatto tree and decided to watch until he was fed. Nearly two hours passed before his mother arrived, and during this interval he was out of my sight for less than five minutes. Occasionally he made a short dart into the air, apparently to catch an insect too small for me to see, but he could have taken little food during his long wait. When finally, at 1:13 P.M., his mother alighted on the feeding perch and called, he briefly vibrated his half-spread wings like a hungry passerine fledgling, repeating this several times with momentary pauses and accompanying each flutter of his wings with a little weak *peep*. Then he flew to his parent and was fed in one installment. Evidently he was still hungry after this brief meal, for he followed his mother to another perch and alighted beside her with open bill, but he received no more on this occasion. On the following morning he was fed more often, four times in three hours, which is about the rate of feeding for each member of a brood of two older nestlings.

While resting on or near the feeding perch, this young Scaly-breast sang so much that I was convinced he was a male. His performance sounded like a very subdued version of an adult male Scaly-breasted Hummingbird's song. Weak, high notes followed each other without much order, forming a sort of medley. Often the notes were so slight that I could not hear them above the roar of the distant river, although the songster was only 20 feet (6 meters) from me. Even when his voice was inaudible and his bill closed, I had no doubt that he was singing, for his strongly distended throat vibrated, its partly erected feathers separated from each other and moving up and down. At such times, I could hear very faint notes when I approached within 2 or 3 yards (1.8 or 2.7 meters) of the hummingbird. This whisper-song often continued for many minutes together. Rarely he probed a flower, sometimes while clinging beside it. When a small black bee hovered around his head, he retreated to a neighboring twig, then soon returned to his preferred perch.

I last saw this juvenile receive food from the parent with a wart on her bill on October 13, when he was sixty-five or sixty-six days old and had been flying for forty-one days. Another Scaly-breast was fed until at least fifty-two days old. Other hummingbirds are nourished by their mothers for comparably long periods: Band-tailed Barbthroats to the age of fifty-six days, Rufous-tailed Hummingbirds to fifty-eight days, Long-billed Starthroats to at least forty-eight days, and White-eared Hummingbirds to no less than forty days. Tropical hummingbirds attend their fledged young about as long as many small passerine birds do.

While attending her young both in and out of the nest, the female with a wart on her bill drove away other birds much larger than herself. She chased a Buff-throated Saltator for about 50 feet (15 meters). One day after her young left the nest, a Blue-diademed Motmot alighted in a small tree near it. The parent darted back and forth, passing close to the motmot but apparently not touching it, at least a dozen times before the intruder flew away, with the hummingbird pursuing it closely until both were beyond view.

In their long breeding season, Scaly-breasted Hummingbirds may nest three times, occasionally laying again in a structure from which young have flown or from which eggs have been lost; but I have not known them to raise more than two broods to nest leaving. One female resumed laying seventeen days after her two nestlings flew from the nest, and later the interval was only eleven days. Of the nineteen nests of known outcome in which at least one egg was laid, ten escaped disaster until the young flew. Of the thirty-seven eggs laid in these nests, twenty-six hatched and eighteen yielded flying young. Thus, 52.6 percent of the nests were successful, and 48.6 percent of the eggs produced fledglings. Some of the Scaly-breasts' larger neighbors did less well. By building nests that blend well with their setting on branches and twigs covered with mosses and lichens, these hummingbirds often escape the predators that destroy so many eggs and nestlings of all kinds.

15. *Our Animal Heritage*

More than a century has elapsed since the publication of Charles Darwin's *The Origin of Species* laid the foundations of our present understanding of the evolution of organic forms and functions. Today practically all serious biologists believe that the more complex organisms arose by gradual modification from simpler, more primitive forms during the course of geologic ages. No alternative explanation of their origin presents a serious challenge to this view. But, despite the firm establishment of the evolutionary theory, there has been widespread emotional reluctance to accept its full consequences, because certain people feel that man is somehow degraded by his derivation from simpler or, as we often say, "lower" forms of life.

Even Alfred Russel Wallace, coauthor with Darwin of the theory that evolution proceeds by the natural selection of usually slight individual variations, believed in his later years that certain attributes of mankind, including our moral sense and more or less hairless skin, could be explained only by assuming that human evolution has been guided by a superior intelligence not operative in the evolution of other branches of the living world. A book that attained great popularity in the mid twentieth century, *Human Destiny*, by Lecomte du Noüy (1947), made the curious distinction between adaptation, the criterion of which is usefulness, and evolution, the criterion of which is liberty. On this view the lineage that culminated in man can alone be said to have evolved; all the others simply became adapted to their environments.

The assumption that a special principle has been at work in the evolution of mankind is fatal to serious, honest thought about the problems of evolution. Either an evolutionary theory must be adequate to account for the origin and present condition of all living things, or it is to be regarded with great suspicion. This does not mean that our present knowledge of evolution permits us to trace in detail the origination of all the features that the living world presents.

Buff-throated Saltator with food for nestlings

Probably every biologist of wide experience is familiar with structures and habits, in animals and plants, for whose origin he lacks an adequate explanation. He attributes his perplexity to the bewildering complexity of the forces at work, the immensity of the periods through which they have acted, and the inadequacy of the fossil record. He does not divide organisms into two or more categories, some of which evolved according to one set of basic principles and some according to wholly different principles. This would be most confusing.

The underlying reason for this recurrent attempt to remove man from the general evolutionary scheme is revealed by the statement of Max Otto (1949) that "the hopelessness about himself into which contemporary man has fallen is reinforced by the belief in his animal ancestry." Although the evidence that man has descended from nonhuman ancestors is too strong to be resisted, some thinkers have supposed that to believe we have been raised above our animal forebears by the operation of a special principle or the guidance of a Superior Intelligence should diminish the shame and despair that darken certain minds when they contemplate man's origin.

Far from causing us to despair about ourselves, the evolutionary view, even in its standard form, should be more encouraging than the older, biblical view that it has been slowly supplanting in the Western mind. To believe that we have arisen from a simpler form of life, however low and brutal it might have been, makes me more hopeful about the ultimate possibilities of myself and my kind than to believe that we fell from the higher state of uncorrupted innocence in which God created our first progenitors. There is certainly no necessary connection between hopelessness about ourselves and the view that mankind is closely allied to other animals. Although they lacked a sound theory of organic evolution, the thinkers of ancient India, like a number of the most eminent of the Greek philosophers, believed that the soul, our very inmost self, might successively inhabit both animal and human bodies, and a more intimate connection between men and other animals is difficult to conceive. Yet these thinkers did not, for this reason, doubt our ability to attain, by appropriate effort, the highest blessedness.

A number of recent writers have expatiated upon the less admirable traits that we have inherited from remote ancestors and share with a wide variety of contemporary animals, including aggressiveness, hostility, the will to dominate, the predatory habit, and sexual drives difficult to control. Sheltered by the protecting ramparts of civilization, some of these attributes or habits have in man become exaggerated to a degree equaled by few other animals, like weeds that grow rankly within the walls of a neglected garden. To counteract this tendency to dwell upon the undesirable components of our animal heritage, we should give equal or greater attention to the elements of this heritage that we can applaud.

In many years devoted to the study of animals in their natural state, I have seen more to make me proud of that close relationship to them which the evolutionary theory posits than to make me ashamed of this affinity. It has become evident to me that many of the human traits which I most admire are exhibited also by birds, mammals, and even "lower" forms of life, and the recognition that these attributes are not restricted to mankind but far more widely diffused through the living world helps bolster an optimism that is often difficult to preserve in these troubled times. Let me briefly enumerate some of the valuable traits which we share with other animals.

Perseverance. Whatever our aim in life, whatever the goal we set for ourselves, we are unlikely to attain it without the ability to persist in the face of the many obstacles that block our advance. All great accomplishments are achieved by perseverance. But this is not an exclusively human trait. We recognize it in the spider that again and again reconstructs her torn web; in the castlebuilder struggling to raise long sticks through obstructing tangles of vines; in the pair of birds who, after the loss of eggs or young, begin afresh to rear a family, repeating this perhaps half a dozen times before they succeed; and, in an extraordinary degree, in the migratory bird who, despite fatigue, must fly on and on over a vast expanse of open water or barren desert until at last it reaches sheltering vegetation. Perseverance is one of the most fundamental attributes of life without which it would never have become what it is. Man seems more prone than most creatures to be discouraged by failure. Organic perseverance is the root of the moral virtue of fortitude.

Courage. Animals sometimes take risks in the pursuit of food, but they could not habitually jeopardize their lives to satisfy their hunger without endangering the existence of their species. It is in the defense of their young that animals display the most amazing disregard of personal safety. Who can withhold admiration from the birdling, scarcely bigger than a man's thumb, who pecks or bites the hand extended to touch her nestlings? Many mammals, too, confront larger and more powerful animals that threaten their young, and even tiny nest-building fish dart at intruders that jeopardize their progeny. The tendency of parent animals to risk their lives in defense of their young is held in check by natural selection, for, considering the great odds that confront them, the species in which devoted parents frequently lost their lives in a fruitless effort to save helpless offspring would soon become extinct. Were it not for this restraining factor, we would doubtless witness even more frequent instances of supreme valor.

Parental devotion. Courage in defense of home and young is only one of the forms that parental devotion takes. It reveals itself even more consistently in the day-to-day task of keeping eggs warm, in the often exhausting labor of feeding the young in foul weather as in fair. The watcher of birds often sees them relinquish a mouthful of food to their nestlings when they show plainly that they themselves are hungry. Monkeys, apes, and some other mammalian parents laboriously carry their young until they grow large and heavy or bear them to safety when danger threatens.

Friendship and love. Certain animals are capable of strong personal attachments, usually to other individuals of their own kind, but sometimes to individuals of other species. Naturalists have recorded many instances of the latter sort, and often it is difficult to point to any material advantage that the animal gains from this close companionship. In many species of permanently resident birds, a mated male and female keep close company, flying and roosting together through the long months when they do not nest and their sexual impulses are dormant. A mated pair of swans or geese remains together for years, even on migratory flights of thousands of miles. Pairs that do not separate after finishing their shared task of rearing their young appear to be held together by a personal attachment that we can compare only to friendship in ourselves.

Cooperativeness. The most remarkable examples of cooperation in nonhuman creatures are provided by the social insects, including termites, ants, bees, and wasps. Since in many of these societies the workers are unable to reproduce and could not long survive apart from their community, we might compare these associations to the interactions of the cells or organs of a living body rather than to the free cooperation of complete individuals. Many vertebrates, from fish to birds and mammals, seek their food in schools, herds, or flocks which at least among birds are often composed of many species. Likewise, gregarious animals often join forces to repel enemies. The most active and appealing mode of cooperation is that often found among birds, especially permanently resident species of mild climates, in which unmated individuals may help the parents feed and otherwise attend their young. Usually these voluntary assistants are older offspring of the same parents who care for younger brothers and sisters. Sometimes the helpers are less closely related to the parents and they may even nourish young of another species.

Joy in living. This is most clearly manifest in play, that is, in activities of no immediate utility prompted by inner urges which appear to give pleasure. We witness play in the frolics and mock battles of young animals of many kinds, in the galloping of well-rested horses about their pasture, in the soaring and diving of large birds in updrafts of air, in the racing of a school of dolphins just ahead of the prow of a steamer going at full speed. Play reveals that the effort merely to keep alive does not drain all of an animal's vitality, but that it enjoys an excess available for spontaneous activities which gratify.

Repose. Many kinds of animals spend much time resting, even by day, when (in purely diurnal creatures) their activity is not inhibited by darkness. When animals rest, they appear to do so thoroughly, not with fidgeting and impatience, as people often do. This capacity for complete repose is certainly a trait of no small value which restless humans, especially in highly industrialized communities, may be in danger of losing.

Appreciation of beauty. As was pointed out in chapter 3 it is difficult to account for the charming songs or elaborately ornate plumage of many birds, the decorated gardens of the bower birds, and similar manifestations unless we attribute to them at least rudimentary aesthetic sensibility. Although not lacking, the indications of an aesthetic sense in other groups of animals are less positive. The sensory and psychic organization that makes us responsive to beauty did not suddenly spring up in us without antecedents but was gradually perfected through a long series of prehuman ancestors, and for no item in our animal heritage should we be more grateful.

Curiosity. Fairly widespread among the more advanced animals is the desire to know what is happening around them or to learn what is in or behind an opaque object. Unfortunately, as in children, curiosity may lead to destructiveness; having no developed technique for examining the object that puzzles it, the ape crudely tears it to pieces. But in such curiosity we detect the first tiny germ of science.

Temperance. Although some animals, especially the smaller birds and mammals, consume, relative to their size, much more food than we do, they need this to support their more rapid metabolism. Free animals cannot afford to be gluttonous, which would be fatal to creatures whose survival depends upon preserving fitness.

Integrity. The activities of a free animal—how it procures its food, migrates, wins a mate, builds a nest, attends its young—compose an integrated system often called its pattern of behavior. A normal animal preserves this pattern as far as it can, even in the most adverse circumstances. Sometimes it succumbs rather than deviate from it, when the human observer sees clearly that by modifying its hereditary procedure it might preserve its life. "How stupid!" he is then likely to exclaim. But the animal who dies rather than change its ways has something in common with the person who forfeits his life rather than abandon his moral or religious principles. We humans need verbally expressed rules of conduct precisely because we are not innately endowed with a pattern of behavior adequate for our guidance. A moral code is the human equivalent of the animal's pattern of behavior; accordingly, to deviate from the latter is analogous to being unfaithful to the former.

It is evident that in animals we detect, in a more or less developed stage, a considerable proportion of the attributes upon which we humans most pride ourselves. We inherited at least their rudiments from our prehuman ancestors; we became what we are by the further elaboration of our animal heritage. Believing that the whole dignity of man resides in his possession of some of these qualities, we often indignantly deny their origin, fearing that to acknowledge it would somehow degrade us. We use all our ingenuity to argue away the obvious similarity between the human trait and the corresponding behavior of the animal. Thus, it is commonly said that a woman's devotion to her child is an expression of maternal love, whereas the animal's attachment to her offspring is parental instinct. In situations where we act from a sense of duty, the animal is said to follow its innate pattern of behavior. In view of our ignorance of the precise relation between our own mental and bodily states on the one hand and of the psychic life of animals on the other hand, these distinctions rest upon a precarious foundation. When close external similarity between a human activity and an animal activity is evident, it is perverse to deny the affinity of the former to the latter.

Primitive clans were proud to trace their descent from their totem animal. But in modern times there has been a persistent conspiracy to vilify animals, in part because we seem thereby to justify our merciless exploitation of them, in part because it is believed that by this means we somehow exalt mankind. But this rests upon confused thinking. If man is higher than the animals, then the more admirable the animals are, the nobler man must be. We do not demonstrate that a mountain is lofty by showing that it is higher than an anthill, but if we can prove that it exceeds Chimborazo or Kilimanjaro, we have made it a high mountain indeed. Similarly, the more that is fine and admirable that we detect in animals, the more reason have we, who believe ourselves to be higher than them, for respecting ourselves.

When I survey the vast array of forms that the animal kingdom has produced, the great beauty of many of them, and the immense diversity of their abilities, I see no reason to be ashamed of my inclusion in a division of the living world that exhibits such marvelous capacity for development in many directions. Such a survey enhances rather than diminishes my estimate of human potentialities. Each of us is what he is at the present moment, and how he came to be what he is neither adds nor subtracts from his physical, intellectual, or moral stature. But knowledge of how we reached our present state influences our estimate of our prospects; and to know that one has arisen from a group of animals so productive of splendid forms and accomplishments as the vertebrates is more conducive to a hopeful outlook than to believe that the human stock has fallen from a higher estate or remained at its present level for countless generations.

It is unfortunate that most people who have some acquaintance with animals know only captive or domesticated individuals, for this tends to give them too low an opinion of animal life and may make them ashamed or resentful of their evolutionary origin. They forget that the common domesticated breeds have for thousands of years been selected for qualities useful to man, such as the capacity to produce flesh or milk or eggs, to bear or haul loads, or to chase and worry other creatures, with little regard for all their other qualities. Not intelligence but docility, not spirit but abjectness, not grace but obesity are required of the majority of domestic animals. All those beautiful and intricate patterns of behavior which so well adjust the free animal to its environment have been distorted or destroyed by generations of subjection to man's will. The wonder is not that creatures who for so many generations have been knocked about and thwarted should so often disgust us but that we should still find so much which is amiable and attractive in them. But to know animal life at its best and estimate it fairly, one must watch free animals in their natural habitats, preferably while they remain unaware of their observer.

It would be perverse to deny that even free animals exhibit, among much that is admirable, certain disagreeable traits. They are capable of selfishness and rage, and they may bully weaker individuals of their own or other species. But what most distresses the sensitive observer of nature is the callous way in which the carnivorous kinds kill, tear, and devour their victims, which may be animals of related species and are often helpless young. Nevertheless, we cannot on this ground refuse to acknowledge our affinity to them without at the same time repudiating our relationship to humans who, with less excuse, slaughter and devour countless animals of the most diverse kinds, tender young no less than the aged and crippled, which may form the mainstay of the diet of wild carnivores. For man, sprung from a vegetarian or omnivorous stock, is neither by structure nor function restricted to a carnivorous diet and could live well without slaughtering animals, whereas predatory beasts would, in most instances, perish if deprived of the prey to which they have become specialized.

Even after giving full weight to the disagreeable traits that some animals exhibit, we have less reason to be ashamed of our place in the animal kingdom than of our frequent failure to make full use of our human capacity for foresight and moral choice. That which most sharply distinguishes us from other animals is our ability to look ahead, to assess competing motives and compare alternative courses and choose according to an ethical standard rather than in blind obedience to appetites or emotions. We have greater need of this faculty than other animals because we have not, like them, an innate pattern of behavior in conforming to which we might safeguard our species, since it has been tested and perfected by a long racial experience. Our morality, too, has roots in our animal heritage, but it has been highly elaborated by a long evolution peculiar to our branch of the vertebrates. Not our affinity to the animals but our much closer kinship to people whose conduct outrages all that is best in humanity is what should make us ashamed.

Our human reason, even at its best, and our highest ideals serve chiefly to modify and direct motives and affections that come to us from our animal ancestors; without them, we would lack incentives to act. We can trace to our animal forebears our staying power, our courage, and at least the germs of our capacity for love and friendship and devotion and our response to beauty. Our reason did not create these attributes; our morality grew out of them instead of producing them. When we awake to the full value of our animal heritage, far from being ashamed of it and wishing to deny it we shall acclaim it as, under the guardianship of our reason and capacity for moral choice, one of the greatest sources of our strength.

We need more pride, but not of the kind that the doctors of the spirit have condemned: not the vainglory that holds one's self better than one's neighbors, one's own race superior to other races; not the egotism that absurdly overrates one's true worth; not the hauteur that disdains to carry a bundle or engage in useful manual labor. What we need is the kind of pride that vetoes any act, however profitable, that might soil our character; the kind that makes a diligent housewife keep her home clean and neat, that compels a craftsman to excel in his craft. After all, to be among the more successful products and in certain respects the highest of two or three billion years of evolution, counting from the genesis of life, is something to be proud of. This is, to be sure, not a personal or even a racial achievement so much as something that was done for us, something that we inherit, as the scion of a distinguished family inherits an honored name. And just as such a descendant of respected ancestors will, if proud of his name, do nothing unworthy of it, so will we, if mindful of our evolutionary heritage, try always to comport ourselves in a manner that dignifies it. Too widespread is the habit of complaining about what we lack and what is wrong with us; we need the pride that appreciates and makes the best of what we have.

16. *A Bird of Stormy Heights*

The high mountains of Costa Rica and western Panama, separated from similar elevations by the lowlands of central Panama to the east and those bordering the Río San Juan and the Lake of Nicaragua on the north, are the home of some fifty species and seventy-five subspecies of birds found nowhere else. For a long while, the most accessible point where many of these birds and associated plants could be seen was Irazú, a huge, sprawling volcano that rises to a height of 11,412 feet (3,480 meters) north of Cartago, the colonial capital of Costa Rica.

Nearly fifty years ago, before the road was paved to the summit, I was riding a horse between pastures and vegetable gardens down the southern face of the then inactive volcano when a flock of starling-sized birds flew across the road and settled in the top of an oak tree. Through my binocular, I scrutinized their slender gray forms with high-peaked crests, yellow under tail coverts, and black tails with white in the middle of the outer feathers and two projecting central plumes. Already familiar with the Gray Silky-Flycatcher of the highlands of Guatemala and Mexico, I recognized them at once as my first members of the more ornate southern species, the Long-tailed Silky-Flycatcher. These two species, with the Phainopepla of southwestern United States and Mexico and the Black and Yellow Silky-Flycatcher, which shares the range of the Long-tailed at high altitudes in Costa Rica and western Panama, comprise the small silky-flycatcher family, the Ptilogonatidae, which accordingly contains only four species in three genera. Sometimes the family is united with the waxwings of northern lands.

During many seasons in Costa Rica, I watched nesting birds at altitudes too low for silky-flycatchers; even the wild mountains where I studied the Resplendent Quetzal were not high enough for these gray wanderers of the cool heights, who rarely descend below 6,000 feet (1,830 meters) and thrive up to timberline at about 11,000 feet (3,355 meters).

Long-tailed Silky-Flycatcher

Finally, in 1963, I found a place to study silky-flycatchers at La Giralda, a large dairy farm at the western end of the massif of extinct Volcán Barva, where Don Rafael Angel Fernández Soto and his wife, Doña Consuelo, founders of the farm, generously lent Pamela and me their big house for nearly five months. From it we looked over the western part of the central plateau toward the Pacific Ocean, which on very clear days was visible in the far distance, while at night the lights of Alajuela and several smaller towns twinkled brightly 3,000 feet (915 meters) below.

The dwelling stood high above the highway that passed over the continental divide in the saddle between Volcán Barva and active Poás, and then descended steeply into the lowlands of northern Costa Rica. The pastures around it were shaded by Alder trees, introduced Mexican Cypress, and, at higher altitudes, by Cornel, Winter's Bark, and several species of myrtles. For better management of the grass, the pastures were divided into small plots by fences. Walking through them to visit birds' nests, I had to open and close many gates if I did not squeeze through the barbed wire. In the numerous ravines that intersected the terrain were patches of ancient and second-growth forest, from less than an acre to several acres in extent, which had been preserved to protect the watershed. To the northwest of the pastures was a long deep valley whose sides were covered with several hundred acres of forest dominated by huge, epiphyte-burdened trees, including many oaks. The steepness of the wooded slopes and the dense undergrowth of tall, canelike bamboos made it extremely difficult to move around and look for birds.

In these woodlands I never saw silky-flycatchers, far-ranging, restless birds of open spaces. They fly in straggling flocks well above the ground, usually above the treetops, whether they travel over the mountain forests or over pastures with scattered trees. Voicing sharp notes and rattles, they flap their wings intermittently to rise and fall in long undulations. Often at the same time they veer from side to side as though in doubt where to go. They alight by preference on the topmost exposed twigs of tall trees, where they perch very upright, with their high crests and long tails appearing elegantly slender and presenting an unmistakable profile against the sky. Even in the gales that blow over the high mountains for hours or days together, the silky-flycatchers choose these exposed perches, where they rest while the boisterous wind ruffles their plumage and twists their tails. Here they remain while the chilling cloud-mist drifts through the trees, dimming their thin figures until they appear insubstantial and ghostlike—then, seeming to dissolve in the mist, they vanish and only their sharp *che chip* betrays their continuing presence. The rigors of the high mountains, with their cold rains and fierce winds and frosty nights in the dry season, seem not to trouble these hardy birds, who are tolerant of most climatic extremes except the heat of the lowlands.

Long-tailed Silky-Flycatchers eat large quantities of both insects and small fruits. I watched them plucking the black berries of the Tree Fuchsia, the orange berries of a mistletoe, and the little fruits of *Eurya theoides*. They are especially fond of the pea-sized orange berries of *Citharexylum*, borne in long, slender, dangling racemes. They gather these and other berries either while perching beside a cluster or clinging to it rather than while on the wing. With their small and apparently weak bills, these birds often have difficulty detaching berries that are somewhat firmly attached. I watched one trying to pick berries of *Citharexylum mocinnii* which appeared by their color to be ripe. Even by giving its head a sideward twist while holding a berry in its little bill it sometimes failed to break it loose. And it dropped two berries after laboriously plucking them, apparently because they slipped from its grasp.

They catch insects in the air, often by long sallies from a high lookout. Some of the most spectacular displays of flycatching that I witnessed were given by parents collecting insects for their nestlings. As the bird, adding insect to insect in its mouth, twisted and turned high in the air with marvelous grace and skill, it often spread its tail, whose projecting central feathers and white and black areas contrasting with the yellow beneath the tail formed a striking pattern. In such aerial foraging, the silky-flycatcher is no less skillful than the Tropical Kingbird, which, like the former, subsists on a mixed diet of insects and small fruits.

Silky-flycatchers are noisy and loquacious but nearly songless. Their most common utterance is the *che chip, che chip*, sharp and dry, which they repeat both while they perch and while they fly. Although primarily a location note, it is also used to express apprehension or alarm; parents whose nestlings appear to be in danger repeat it incessantly. When a flock of silky-flycatchers perches in a high treetop, all calling *che chip*, the effect produced is that of rattling loose pebbles in a box.

As it takes wing and from time to time in the course of its flight, the silky-flycatcher often voices a long-drawn-out *che-e-e-e-e*, a rattling or clicking sound. Although frequently dry and harsh, this utterance sometimes becomes clear and metallic. Occasionally the flying silky-flycatcher almost achieves a brief, clear trill, which reminded me of the flight sounds of the Turquoise Cotinga. Bell-like or tinkling notes are sometimes given, especially by birds on the wing.

One flight call sounded like *re-er-re-re*, clear and bell-like. In allusion to these sounds, the silky-flycatcher was called *el timbre* at La Giralda, but I do not know how widely this name is applied to the bird.

Rarely I heard a male silky-flycatcher, perching in sunshine on a high, exposed twig while his mate incubated, singing with low, lisping notes that were scarcely audible at a distance of 25 yards (23 meters). I might have missed them if my attention had not been drawn by slight movements of the songster's bill and throat. The low, soft notes of this whisper-song were punctuated by louder *che-chip*'s. Once I heard a silky-flycatcher sing in this manner while he flew. I suspected that silky-flycatchers, like a number of other birds with slight vocal gifts, might perform more loudly and persistently at daybreak; but if these birds have a dawn song I failed to discover it.

Some birds, including many that dwell in the undergrowth of tropical forests, are so solitary that they are usually seen alone or at most with a mate; others, including Cedar Waxwings and many kinds of parrots, have achieved a high degree of social integration and fly in compact flocks. Silky-flycatchers fall into neither of these two categories; they are neither solitary nor given to performing coordinated group movements. Although they seek the company of their own kind, they refuse to surrender their independence of movement. One cannot watch them for long without concluding that their life is a continual compromise between gregarious and individualistic impulses.

Early on an afternoon at the end of February beneath a cloudy sky, I watched a number of silky-flycatchers, at one time sixteen or more, resting in the top of a tall Alder tree in a pasture and keeping up a constant rattle of sharp notes. Frequently they shifted their positions in the treetop, and at times one flew at another, who without resistance relinquished its perch to the claimant. The flycatchers were continually arriving in the treetop or leaving it, coming and going in all directions in parties of one to seven. Sometimes a number would start off in a loose flock, from which some turned back to the treetop while others continued onward. The latter, instead of remaining together, would often diverge to follow separate courses until out of sight. These and many similar observations showed how poorly developed the flocking habit is in these birds. Their flocks are even looser than those of toucans, who usually fly in the same direction as their companions, although they proceed one by one in a straggling party.

One evening in April, as I ended an all-day watch at a nest of an incubating silky-flycatcher, four of these birds flew close together down the mountainside through the mist, passing near the nest. As I walked down through the pastures, I noticed nine silky-flycatchers resting close together on the top of a tall dead tree. After remaining motionless for some minutes, they all flew off together in a compact flock, apparently to roost in company. They behaved more like Cedar Waxwings than silky flycatchers, seeming to be drawn closer together in the twilight than in full daylight. I did not learn where silky-flycatchers slept. In the evenings, the mates of incubating or brooding females would fly off beyond view, sometimes with other silky-flycatchers, leaving the females alone in the nest.

The mountain slopes between 6,500 and 7,500 feet (1,980 and 2,285 meters) above sea level on which I studied the silky-flycatchers were close enough to the continental divide to feel the influence of both the Pacific Ocean and the Caribbean Sea. From our arrival in late February until early in April, the prevailing winds came from the Caribbean, across the northern lowlands of Costa Rica and the crest of the range to our north. Sometimes these northeasterly winds blew steadily all day, occasionally attaining such force that they broke branches from trees and made bird watching unprofitable and dangerous. Since they had dropped most of their moisture while rising to the continental divide and were now blowing downward, they did not bring much rain but chiefly clouds and drizzle, which they drove through the trees, sometimes continuing for a day or two.

Sometimes, veering more to the east, the wind brought ash from erupting Irazú, 20 miles (32 kilometers) away. Early in the morning of March 22, a strong easterly wind bore enough fine volcanic cinder to obscure the sun and cover all the vegetation with a thin layer of gray dust which persisted, especially in ravines sheltered from the wind, until washed off by the hard rains that came two weeks later. Until our departure in July, there were occasional days when ash fell, although not nearly so much as on the central plateau southwest of the erupting volcano.

It was difficult to assess the effect of this volcanic cinder on the silky-flycatchers and other birds. A Black-cheeked Warbler, who had been building very actively on March 21, did not resume work on the following morning when ash darkened the sky; thereafter, her nest was abandoned. But a Flame-throated Warbler, who had started her nest ten days earlier, continued to attend it and raised her brood. For most of the local birds, the nesting season began later than I expected, possibly because of the ash. On a visit to the more heavily affected central plateau in February of the following year, I was surprised by the number of birds of a variety of species who were living, apparently in good health, amid vegetation that for over ten months had been almost constantly dusted with volcanic ash, which now in the dry season lay heavily on all the foliage.

Despite days of strong wind, drizzle, and falling cinders, we enjoyed many beautiful, calm, clear days at La Giralda in March. The wet season arrived on the afternoon of April 5 with a heavy rain, the first soaking downpour in a month. Westerly breezes now became common; although they were rarely as hard as the northeasterly winds, they drove in the rain-laden clouds from the Pacific. As the rainy season advanced, sunny days became fewer. Often, after a few hours of intermittent sunshine in the early morning, we had mist and drizzle for the remainder of the day, with perhaps a torrential downpour in the afternoon. Sometimes the wind shifted to the north or northeast and blew violently down the mountainside, driving the clouds.

On April 1, I found my first pair of silky-flycatchers starting a nest, but I did not discover another building pair until after the middle of the month. The last two of the eighteen nests that I watched held half-grown nestlings in early July, when we left the farm. Thus, the silky-flycatchers, like most of their feathered neighbors, bred in the early part of the long rainy season, when plants that had flowered in the drier months yielded many berries and insects were abundant, amid mist and chilling rain and gales so strong that I expected their nests to be torn from the trees.

Breeding pairs of silky-flycatchers were not uniformly distributed over the hundreds of acres of mountain pastures where scattered trees offered sites for their nests but were grouped in small, widely distant colonies, each containing from two to four nests. The closest nests were only 65 feet (20 meters) apart, but more often those of a colony were separated by 150 to 225 feet (46 to 69 meters). Each pair defended a poorly defined area extending around its nest site for a distance of about 75 to 100 feet (23 to 30 meters). Both members of the resident pair flew at trespassers of either sex, who always retreated without resistance, so that I never saw a clash between them. When nests were exceptionally close together, one of them might be screened by dense foliage so that its attendants could come and go without attracting the attention of their nearest neighbors.

In choosing the sites of their nests, silky-flycatchers show the same poorly integrated mixture of individualistic and social tendencies that we noticed in their flocking habits. They like to build near one another, but their colonies are of the loosest sort, and some pairs nest far from others. In addition to providing the company that silky-flycatchers appear to enjoy in moderation, the propinquity of nests permits them to warn their neighbors of the approach of enemies and to unite in chasing them.

The silky-flycatcher's bulky nest needs a firm support because it becomes fairly heavy when soaked with rain. Often it is placed in a fork of an erect or ascending branch, usually well out from the trunk, but it may be on a stout horizontal branch at or near the bottom of the tree's crown at a point where upright twigs support it on the sides. In height, nests that I found ranged from 6 to about 60 feet (1.8 to 18 meters), but most were from 10 to 20 feet (3 to 6 meters) above the ground.

The male silky-flycatcher—who could be distinguished from his mate by his pale yellowish head and neck, by his bluish slate color rather than olivaceous gray body, and usually also by his longer central tail feathers—worked as hard at building the nest as she did. The chief constituent of all the nests was a profusely branched, pale gray beard-lichen (*Usnea*) that grew abundantly on almost every tree of these cloud-bathed heights. This material, like everything else that went into the nests, was pulled directly from the trees rather than picked up from the ground, on which silky-flycatchers seem never voluntarily to alight. To pluck these lichens was no easy task for these weak-billed birds, who often, after tugging strenuously at them, either while perching or while hovering beneath them, were content to fly to their nests with scarcely visible fragments in their bills. Minor ingredients of the structures were small twigs and pieces of dry inflorescences, foliose lichens, and cobweb or caterpillar's silk, which was freely used to bind the lichens together.

An actively building pair brought material to their nest forty-two times in one hour and twenty-two times in the following hour. Each partner arranged in the nest whatever it brought, scarcely ever passing its contribution to the mate when it found the latter sitting there, shaping the structure, as some birds habitually do, thereby saving time. To compact the bushy lichens into a dense, feltlike mass required much vigorous work which seemed to be done chiefly with the feet. Sitting in the nest, the builder of either sex depressed its breast and apparently kicked or kneaded with its feet, which of course were invisible to the watcher on the ground. After doing this for a short while, the bird would turn sideward and repeat the performance until, alternately turning and depressing its breast while it appeared to kick or knead, it had in some instances made a complete turn; once it rotated, little by little, through about 450 degrees. The builder might turn either to the right or to the left. While engaged in shaping the nest, the silky-flycatcher often lowered its crest, sometimes laying these feathers quite flat so that they projected only a little at the back of the head. As the bird pressed its breast deep into the nest, one or both wings were raised slightly above its back, with the longer plumes somewhat spread. Its bill was used chiefly to arrange materials on the outside of the nest and to spread cobwebs over the rim.

Not only did the male silky-flycatcher take his full share in the task of building, he also fed his building mate, usually not while the birds were actively working but while they rested from their labor in nearby trees. Then the male might fly out, catch one or more insects in the air, and pass them to his partner. A minute or two later, he might repeat his gift. At one nest, however, I saw the male feed his mate twice in the course of two hours while they were busily building. On one occasion, he offered her food as she approached the nest with material in her bill. I did not see clearly what she did with this inedible stuff, but the vigorous swallowing movements that she made suggested that she ate it along with the food. The nuptial feeding begun during nest construction, if not earlier, continues while the female lays and incubates.

The finished nest is a beautiful structure, unlike that of any other bird that I have seen. The broad, open cup is composed almost wholly of a single kind of light gray beard-lichen, compacted into a firm, thick, resilient fabric. The few fine twigs, dry flower stalks, and similar pieces mixed with the lichen are hardly noticeable. Pieces of foliaceous lichens, gray on the upper surface and brown or blackish below, seem always to be present and on some nests are rather liberally attached over the outer surface, as on certain hummingbirds' nests. The cohesion of the structure owes more to the felting or intertangling of the innumerable fine branches of the predominant beard-lichen than to the few tufts of cocoon silk mixed in the mass or applied to the rim, which is broad and neatly rounded. Unlike most birds' nests, this has no special lining; the inner surface of the cup is composed wholly of the same lichens that make up its bulk. As long as a nest is in use, this surface is as hard, smooth, and firm as though it were coated with plaster; but in an abandoned nest, alternate wetting and drying roughens it. When thoroughly soaked with rain, the nest becomes very soft and tends to lose its shape. Three nests collected after predators had taken unhatched eggs measured 4½ to 5 inches (11.5 to 13 centimeters) in overall diameter by 2 to 2½ inches (5 to 6.5 centimeters) high. The cavity was 2¼ to 2½ inches in diameter by 1⅝ inches deep (5.5 to 6.5 by 4 centimeters).

The silky-flycatchers' nests were often situated where little or no foliage concealed them, sometimes on dead branches. Usually, however, they were placed amid a profuse growth of the beard-lichen of which they were composed, which frequently hung in long streamers around them. Thus, they blended so well with their setting that they were easily overlooked, although the practiced eye could often pick them out by their solidity. And Brown Jays, who preyed on the eggs and nestlings, seemed to have little difficulty finding them.

From two to four days after the nest appeared to be finished, the female laid her first egg, early in the morning. Considerably later on the following morning, she deposited the second egg. While she sat in the nest to lay, she received insects from her mate. In no case did I find more than two eggs, whose pale gray ground color, almost matching that of the lichens on which they rested, was blotched and spotted with shades of dark brown and pale lilac. The male, who had often sat in the nest while building it, now no longer entered it, leaving the task of warming the eggs to the female alone. She started to incubate, with less than full constancy, on the day she laid her first egg. Although often a restless sitter, flying from her nest from fifteen to forty-five times in the course of a day, usually she hurried back after a brief recess, so that she achieved a constancy remarkably high for so small a bird. Five females, each of whom I watched for an entire day, were on their eggs for 81 to 87 percent of their active period of ten to eleven hours, counting from their first departure rather late in the morning until their last return early in the evening (Skutch 1965).

While the female sits in her shallow nest, much of her body rises above the rim. Her long tail, held almost horizontally, projects far outward. Incubating females differ in their treatment of their crests, some holding them more or less flat much of the time, others keeping them more than half raised. As night approached, however, all that I watched laid their crests quite flat, greatly changing their appearance. At the same time, they fluffed out the feathers of their breasts and flanks until they spread sideward beyond their folded wings. I would find them again at dawn still in this posture. When rain fell, they also sat with flattened crests.

While the female incubates, her mate spends much time resting on a particular exposed perch, often a dead branch, in the top of a neighboring tree or sometimes in the nest tree itself, if its crown spreads widely. From this lookout he makes sallies to chase away intruding birds of his own or other kinds and to catch insects, some of which he takes to his partner. Frequently he calls *che chip* but he rarely sings, and then in a scarcely audible whisper. When the female flies off to gather berries, he may either accompany her or remain on his perch. When she returns, he sometimes escorts her almost to the nest, but often he neglects this courtesy. For long intervals, he is out of sight.

The high constancy of the incubating female silky-flycatcher does not appear to depend upon the food that she receives from her mate and the consequent reduction of the time that she devotes to foraging. Males fed their incubating partners more or less frequently early in the season but scarcely at all late in the season. Nevertheless, females who received little or no food from their consorts spent as much time on their eggs as those who were fed often. The importance of the silky-flycatcher's nuptial feeding appears to be in maintaining the pair bond and keeping the male informed of happenings at the nest, rather than in substantially reducing the time that the incubating female devotes to foraging, as in goldfinches and other birds. I believe that the female silky-flycatchers whom I watched at their nests could in a few minutes have collected as much food as their mates gave them in a day.

Males fed their mates either on or off the nest. Male 4, who gave his partner food more often than any other, thirteen times in a day, was seen to feed her only at the nest, either in the midst of a session or as she was about to resume sitting after an excursion. If she flew from the nest as he arrived to feed her, he would wait beside it until she returned a few minutes later. Male 3 fed his mate four times at the nest and four times after she had flown from it. All seven of the feedings that I credited to male 2 were at the nest. Possibly all these males gave their partners additional food while they foraged together beyond view. Occasionally the male gave his mate a berry, but much more often he brought her insects, which he might carry in his distended throat as well as his bill and pass to her in two or three portions. Once a female quivered her wings like a fledgling while her mate fed her as they perched side by side in a tree. Usually, however, the female received the offering undemonstratively.

Parent silky-flycatchers seemed to anticipate their nestlings. Sometimes a male went to the nest while his mate was absent, as doubtless he was well aware, and while awaiting her return lowered his head into the bowl as though inspecting the eggs or possibly offering food to them. When the female arrived, she received it. While a female sat in the nest to lay her second egg, her mate brought her some insects. Holding them in her bill, she rose up and seemed to offer them to her first egg, then settled down in the nest and ate the insects. After taking several insects from her mate while she incubated, a different female rose up with one in her bill and, lowering her head, repeated low, throaty notes, as though coaxing the eggs to take food. Another female returned from an excursion holding in her bill an insect which she appeared to offer to eggs still far from hatching. Similar cases of anticipating the nestlings have been recorded of male and female birds of many species.

As already told, even in the early part of the rainy season when silky-flycatchers were nesting the wind sometimes shifted to the northeast or north and blew strongly. One of these windstorms happened to arise while I watched one of the latest nests toward the end of June. Throughout the forenoon, bright sunshine had alternated with intervals of cloudiness and brief flurries of fine drizzle. Soon after the female's return to her eggs at midday, rain began and continued for about an hour, varying from a drizzle to a moderately heavy shower, driven by a northerly wind that frequently became violent while thunder rolled loudly in the distance. I tried to keep dry by pointing my umbrella into the gale, but a sudden boisterous reverse gust turned it inside out. After the rain stopped, the wind continued to blow strongly until late in the afternoon.

Early in the rainstorm, the silky-flycatcher incubating before me had spells of shaking and trembling, especially noticeable in the movements of her wings and tail. After a while, these spells ceased. Through most of the storm, she sat in her nest almost transversely to the wind's prevailing direction. Although her long, exposed tail, presented edgewise to the wind, remained almost straight, the slender projecting ends of the two central feathers were blown strongly to her right. She appeared to have a hook on the end of her tail! While the wind blew most fiercely, she was pressed against the leeward side of her nest; she appeared to be holding on with her feet and straining to maintain her position in the cup. For two hours and sixteen minutes she stuck resolutely to her post, achieving the longest diurnal session of an incubating silky-flycatcher that I timed. At half past two she left during a lull in the gale for a recess of only six minutes, from which she returned as the wind increased in intensity. Throughout the day, she was absent from her eggs only seventy-seven minutes. This was much less time than any other female had taken for foraging in a day, and I doubted whether she could incubate so constantly for several days in succession. The sideward bend of the tips of her central tail feathers persisted until the next day.

From the point where I sat to watch this nest, I could see a nearby nest when the wind blew aside foliage that obstructed my view. The female of this nest also sat continuously through the worst of the storm for about as long as her neighbor with the bent tail feathers. She sat facing into the wind and was wildly tossed on her long horizontal bough. When the wind abated temporarily, both females left at the same time, and they returned together when the wind's velocity increased again. By such steadfast sitting through storms that threaten to tear them from their nests or their nests from the trees, silky-flycatchers succeed in rearing their broods in the frequently severe weather of the stormy heights where they dwell.

From the laying of the second egg to the hatching of the second nestling, sixteen days elapsed at one nest and seventeen days at another, where the female sat less constantly. When I raised a mirror above a silky-flycatcher's nest in which an egg had just hatched, I was immediately struck by the unique appearance of the nestling, not quite like that of any other that I had seen. When, after bringing a ladder, I took it in my hand, the impression of its uniqueness persisted. Its natal down consisted of short, compact tufts that were nearly white and only a few millimeters long. These tufts were arranged in narrow rows between which were large areas of dark bare skin. On top of the nestling's head, the downy tufts formed a wreath or open crown with a dark bald spot in the center. A line of tufts ran along the middle of the back and rump. On either flank was a short row of similar tufts. A line of them ran along each side of the fat abdomen. There were a large patch of down on each wing and tiny tufts on the sides of the head and the thighs. By breaking the nestling's dark surface into irregular areas, the lines of down made it more difficult to recognize. At a distance of a few yards, I saw the nestling as a cluster of dark patches on the nest's light gray floor. Its bill was very short and relatively broad. The inside of its mouth was flesh-colored centrally with a narrow margin of orange. Yellow flanges projected from the mouth's corners. The nestling's legs and toes were pink.

These nestlings were brooded only by their mother, who ceased to cover them on clear mornings when they were about two weeks old, although she continued to shield them from the frequent rains and to cover them by night until they left the nest. Both parents fed the young, at first almost wholly with insects caught on the wing but with increasing quantities of berries as the days passed. Usually a number of items were brought at one time, held in the mouth and throat as well as the bill, and delivered to the nestlings in installments. As many as seven berries might be brought at one time. Just as, while they built, males had placed their contributions directly in the nest instead of passing them to their partner, so now they preferred to feed the nestlings from their own bills. During the nestlings' first days, when some females brooded them continuously for long intervals, making them inaccessible to the other parent, he might perch in a neighboring tree, with mystifying skill catching insects in a mouth from which they already overflowed and holding his collection for seemingly interminable periods. When finally the mother left and often not until she returned with food for the nestlings, he at last flew up to give the young a large meal. Other males would approach the brooding female with a billful of food, pass part of it to her, and then, when she rose up to feed the nestlings beneath her, thereby exposing them, he would deliver the remainder directly to them. In twenty hours of watching, distributed throughout the nestling period, one male brought food 149 times, his mate 110 times. Even after she ceased to brood, this male fed the nestlings more often than she did. At other nests, especially toward the end of the breeding season, females brought by far the greater part of the nestlings' food. The number of feedings per hour varied widely from nest to nest, apparently being fewer when each meal was bigger.

From the hatching of the young until they flew, all their droppings were swallowed by the parents after they had fed them. With the exception of a single shell from which a chick had just escaped, I never saw a parent carry anything from a nest in its bill. The reason for this appeared to be that the droppings were not enclosed in a firm gelatinous sac that made them easy to carry, as in many other passerine birds, but were loose and readily fell apart. Silky-flycatchers defecated more freely than most nestlings, as was very evident when I handled them, and their feces after the first few days contained many seeds. After the nestlings were older and had grown black tails 1 inch (2.6 centimeters) or more long, they often wagged them from side to side after receiving a meal, a signal to the parents that droppings were about to be voided. The droppings of these older nestlings, whose hindparts often projected beyond the nest's rim, sometimes escaped the parent and fell downward. When this occurred, the parent darted down in pursuit and often overtook the falling particles if branches did not interfere. I never saw adult or young silky-flycatchers regurgitate indigestible seeds or the hard parts of insects, as American flycatchers and many other birds do.

Although the parents' efforts to keep the nest clean seemed to be handicapped by the absence of a tough sac enclosing the nestlings' droppings, they succeeded remarkably well. The nests of living lichens from which young had just departed, after more than three weeks of occupancy, were as fresh and clean as when newly made. Few nests that I have seen showed so little trace of the brood that was reared in them.

Parents not only chased other silky-flycatchers from near their nests, but they sent off visiting birds of other species as well. Those that I most frequently saw them expel were Mountain Thrushes, much larger than themselves, and Mountain Elaenias, who are much smaller. These harmless little birds were everywhere in the pastures; occasionally they built their nests near those of silky-flycatchers and often they foraged even closer. Although sometimes mildly chased, they were mostly ignored. Resident Flame-throated Warblers and wintering Wilson's Warblers were occasionally driven from a nest tree.

On a morning in early June, after the young in a high nest were feathered, a Prong-billed Barbet flew into their nest tree, attracted by the small orange berries of a mistletoe that parasitized it. Both parents at once vigorously attacked the brown visitor, darting at it and apparently striking it. The barbet did not receive this onslaught passively but tried to defend itself, which, with its short thick bill that can carve into fairly sound wood, it did most effectively. It seized a silky-flycatcher by a toe and held it dangling until the furious attack of the victim's mate effected its release. Then the barbet flew into a clump of mistletoe only a few feet from the nest where the parents continued their onslaught. The barbet tore a big bunch of feathers from the body of one. Taking the offensive, it darted at its persecutors. Their attacks continuing, it seized one by a leg and held it until barbet and silky-flycatcher fell downward together. While they were falling, the barbet released the leg and then flew away. While this fight was in progress, I heard high, soft notes that apparently came from the silky-flycatchers rather than from the barbet.

The barbet's powerful, three-pronged bill can draw blood from the hand that incautiously grasps the bird, and I feared that it had injured, perhaps crippled, the delicately built silky-flycatchers whose toe and leg it had seized. I was pleased to see both parents fly up and perch normally after the intruder left. They also caught flying insects without difficulty, and the male soon resumed feeding the nestlings. Several times he went to eat the orange mistletoe berries that had attracted the barbet, although for some days I had not seen him take an interest in them.

The female silky-flycatcher seemed to have come off worse than her mate. While he fed the nestlings, she perched much, resting her abdomen against the branch, as though to take the weight off her legs. After a while, however, she took food to the nestlings, then brooded them briefly. Next day, she was attending the young, showing no ill effects from the previous day's encounter. A few days later, a barbet who flew into a neighboring tree fled to the forest as the parent silky-flycatchers approached.

I never saw a silky-flycatcher simulate injury, an omission to be expected in so arboreal a bird. Nor did any of them attack or make feints of attack when I touched their eggs or young, although birds no larger than they have occasionally struck or pecked me in these circumstances. While I was capturing a nestling that fell to the ground, the parents flew close around and alighted on low branches a yard or two from me, uttering shrill cries of anguish or alarm such as I had not heard before. When I examined half-grown young in another nest, the parents flew around quite close to me with spread tails that displayed the white areas on the outer feathers. While building, incubating, or attending newly hatched young, silky-flycatchers seemed nearly or quite indifferent to my presence, so that I could watch all their activities while seated without concealment at a convenient distance. As their nestlings became feathered, they grew increasingly wary and refused to approach their nest in front of me, even after I had sat for a long while a good way off. An outstanding exception was the pair whose high nest I found soon after it was started and watched periodically until the young flew. These parents became accustomed to my presence and seemed never to be disturbed by it.

When the nestlings were about two weeks old, the upper surfaces of their bodies were fairly well covered by expanding feathers, at least when their wings were folded, and they promised to resemble their mother. Although one expects songbirds no larger than the silky-flycatcher to leave the nest when about two weeks old, day after day these grayish nestlings lingered in their gray nest, with the tufts of white natal down, still liberally sprinkled over their upperparts, breaking the smoothness of their fresh new plumage. Their black wing plumes expanded and their black tails, daily becoming longer, stuck up prominently above their nest's low rim. Soon a crest became evident on the top of each head. Nevertheless, these young birds were slow to acquire the power of flight. When I was about to remove eighteen-day-old nestlings for photography, one jumped from its nest. Still unable to fly, it fell, clung to a twig, then fluttered to the pasture grass beneath the nest. Here it tried to escape, but it hopped so weakly that it was easily caught. Thereafter, it rested quietly wherever I placed it.

When the young in the highest nest were sixteen or seventeen days old, I first saw one stand on the rim, preening, for a short while. Six days later, in bright afternoon sunshine, I found these nestlings very lively, preening and flapping their wings, sometimes while perching on a twig beside the nest, to which they soon returned. On the following morning, they repeatedly ventured beyond their nest, sidling out along the more horizontal of the branches that supported it for a few inches but always returning home after a minute or two. As the sunny morning advanced, these bold excursions grew longer until the young were 6 or 8 inches (15 or 20 centimeters) from their nest. At first they ventured forth singly, but by midmorning both were perching beyond the nest. Soon both rested in it once more. A little later, one of the young hopped and fluttered up an ascending branch until it was about 1 foot (30 centimeters) from the nest. From this point it half jumped, half flew to a more horizontal branch, along which it promptly returned home. Now, by one of those sudden changes of weather frequent in high mountains, a dense cloud drifted in to envelop the nest tree. Soon it was raining hard, and the nestlings were brooded almost continuously by their mother.

At seven o'clock on the following day, June 11, these nestlings were resting quietly in their cup of lichens in bright morning sunshine. Soon becoming active, one of them hopped and flew from branch to branch within a radius of 2 or 3 feet (.6 or .9 meter) of its starting point, making a circuitous journey that brought it back to the nest, where the other rested. Presently one young silky-flycatcher, then the other, left the nest again and started to explore the crown of the nest tree, flying from twig to twig until one of them reached the very top of the tree, a yard above its birthplace. They preened and scratched themselves, pecked at the foliage and bark, and made flights of a yard or more through the open crown. They had a strong tendency to keep together.

When the fledglings' father returned and found them outside the nest, he flew from the isolated nest tree to one that stood in a row along the nearby fence between two pastures. Five times in rapid succession, he returned to the nest tree and repeated the 50-foot (15-meter) flight to the other tree, apparently trying to lead the fledglings to denser foliage, but they would not follow him. Presently he fed one of them in the nest tree. A minute later, the female took food to the empty nest, looked in as though expecting to find the young birds there, then carried the food to the same fledgling that the male had fed. She went off for more food, again took it to the empty nest, then fed the other fledgling.

After their short flights, the young silky-flycatchers alighted with ease. They scratched their heads by raising a foot inside the relaxed wing of the same side, just as the adults did. The smaller of the fledglings repeatedly sidled up to the larger one and pecked gently at its plumage. Unlike many birds who have just left the nest, these young silky-flycatchers were in no hurry to seek cover. An hour after they hopped from the nest, they were resting close together in the exposed crown of the nest tree where their parents fed them. For another hour they perched almost in the same spot, sometimes a few inches apart, all in the cold gray mist that now covered the mountain. Here I left them for a while. When I returned later in the morning, one fledgling had vanished, but the other was still resting in almost the same place. When the parents called *che chip* from neighboring trees, it answered with a low *chip*. When a pair of harmless Golden-browed Chlorophonias foraged near the fledgling, its mother chased these small, brilliant tanagers away.

Shortly before noon a shower began, and I watched to see whether the fledgling would return to its nest to be brooded, as with its sibling it had been on the preceding day. For the next twenty minutes, it rested in almost the same spot, while the rain grew harder and the parents perched in nearby trees. The young silky-flycatcher showed no inclination to return to the nest only a few yards away. At noon I left it on its exposed perch in the treetop in the downpour. In the last four hours it had moved very little.

I wondered how the newly emerged fledglings survived the hard, cold rain which fell most of that afternoon and far into the night. Next morning I could not find them; but the parents were carrying food into some trees with dense foliage, where doubtless the young were hiding. They had left the nest when at least twenty-four days of age and possibly one day older. From another nest, the two young departed just twenty-five days after the younger of them hatched.

A noteworthy feature of the young silky-flycatchers' behavior during their last days in the nest was the repeated, increasingly long excursions which finally led to its abandonment. Most fledglings that I have watched did not return to the nest after first severing contact with it. Although certain ovenbirds, cotingas, American flycatchers, and swallows that breed in holes or closed constructions remain in them as long or somewhat longer, the silky-flycatcher's nestling period of twenty-five days is, for a small passerine bird, amazingly prolonged. Such a long period in an exposed nest and the fledglings' failure promptly to take cover after leaving it might be disastrous in a region where predators, especially hawks, were more abundant than I found them on this mountain. On the other hand, to remain in the nest and be brooded when cold rain falls is certainly an advantage for young birds that are reared at high altitudes in the wet season.

At La Giralda I found eighteen silky-flycatchers' nests, in all of which, with one possible exception, eggs were laid. In two of these nests, young were still present when we left the farm in early July. This leaves fifteen nests of known outcome, of which four were successful, producing eight fledglings. From eleven nests, eggs or nestlings not more than a few days old vanished. The success of this small sample was, therefore, 27 percent. Since some of these nests contained eggs or even nestlings when found, they had already survived some of the hazards to which nests are exposed from the day the first egg is laid. Of the eight nests found before eggs were laid, two were successful and one held week-old young when I last saw it.

Here in the high mountains, nest-robbing animals were less abundant than in lowland tropical forests. I did not find a single snake during nearly five months at La Giralda. After the departure of migratory hawks in March, the only raptor that I saw was a single Red-tailed Hawk. Emerald Toucanets, who often plunder nests in the mountain forests, were not seen to molest those of silky-flycatchers. The case was otherwise with Brown Jays, whose approach to nests stirred silky-flycatchers to the highest pitch of excitement and spirited pursuits of the intruders. After jays had passed through one of the flycatchers' nesting colonies, I found fresh yolk from an egg that had just been taken before I could chase them away and save the other egg. The farm's ranger told me that he saw a piá-piá, as he called the Brown Jay, carrying a nestling silky-flycatcher with adults in pursuit. These jays, open-country birds more abundant at low altitudes, were here at 7,400 feet (2,255 meters), near their upper limit, which apparently they had reached in relatively recent times after the destruction of the forests. Probably when Long-tailed Silky-Flycatchers evolved their present nesting habits they had no contact with Brown Jays, whose keen eyes detect the nests of gray lichens despite their concealing coloration. Less tolerant of the loss of part of their clutch than certain other birds, silky-flycatchers at two nests abandoned their remaining egg after predators had taken the other.

Although small birds of many other kinds share the cool, damp, epiphyte-laden trees of the high mountains with the Long-tailed Silky-Flycatchers, scarcely any other so exposes itself to the gales that drive cold gray clouds across them. They seem to delight in their ability to withstand all the sudden changes of mountain weather. Since they are not dependent upon unbroken forests but can thrive where only scattered trees and berry-laden shrubs remain, they may long continue to enliven the stormy heights of Costa Rica and western Panama.

17. Musings at Birds' Nests

The most charming of feathered creatures would not be half so interesting and lovable without their nests. These structures that they skillfully make and faithfully attend are so much part of themselves that without watching them at their nests we cannot know them well.

What a fascinating array of nests birds build! In size they range from tiny structures that would fit into a child's hand to the huge platforms able to hold a man that eagles make. Most nests, from the downy chalices of hummingbirds to the broad bowls of jays and crows, are open above, but many birds prefer a structure that gives better protection to the eggs, nestlings, and incubating or brooding parents. By providing a cup or bowl with a domed roof, leaving a round doorway in the side, birds gain this protection most readily. More elaborate are the hanging nests, attached at the top to a twig or branch, of which birds build a great variety, including long, neatly woven pouches open at the top, retort-shaped structures entered through a downwardly directed tube, or others of the most varied forms with a doorway in the side. Laborious birds build elaborate castles of interlaced twigs, entered through a narrow passageway or containing many chambers, one above another. Other birds burrow into the ground or carve chambers in wood or hard termitaries. To build these so varied structures, birds gather the most diverse materials: straws, stems of herbaceous plants, leaves green or dry, stiff twigs, fine vegetable fibers, seed down, clay, moss capsules, mammalian hair, feathers, spider webs, bits of paper, snakeskin—almost anything that can be shaped into the desired form. They use skills that we associate with weavers, felt makers, masons, miners, carpenters. A collection of birds' nests is no less interesting than a collection of sea shells, but it needs more space and is more difficult to preserve because the materials of nests are mostly less durable than mollusks' shells.

Although empty nests may make an absorbing study, they do not half so richly reward our attention as nests attended by their makers. Because birds have so many enemies, they tend to be elusive creatures; some are so shy or live amid such dense, concealing vegetation that we deem ourselves fortunate to catch a fleeting glimpse of them. Sighting them as they fly by, flit through the trees, or sing, which satisfies many bird watchers, hardly reveals their temperaments or what they are capable of doing. Complex patterns of behavior, programmed in their genes and latent much of the time, are displayed only on appropriate occasions, especially at their nests. When we have found a bird's nest, we are no longer dependent upon chance and often exasperatingly fragmentary views of the bird in the bush. We have established a rendezvous or trysting place where, if all goes well, we can be certain of meeting it at times convenient to us, for weeks or months, according to the duration of the nesting cycle. Watching a bird at its nest is like being admitted to the home of a person we had met only in public places and at last coming to know him intimately. And while we sit, hidden in a blind or otherwise if the parents are shy, contemplating the intricate details of hatching the eggs and rearing the nestlings, we have leisure to meditate upon the significance of what we see.

When we remember that, exposed to lurking predators and other hazards, the nest is as likely—in tropical forests much more likely—to fail as to succeed in its purpose of rearing a brood of fledglings, to build it, lay eggs, and spend patient hours incubating them seems an expression of faith implicit if not fully conscious. For faith, commonly extolled as a religious virtue, is far more than believing what cannot be proved, more than the *credo quia absurdum* (I believe because it is absurd), as Tertullian and churchmen after him declared. Faith is the fortitude to act as if certain unprovable propositions were well established, the capacity to labor without assurance of reward, to undertake what we are not certain that we can accomplish, to create what we are not sure that we shall enjoy. Without such faith, some of man's grandest achievements would never have been undertaken, and even commonplace tasks would often be left undone. We set forth on our journeys without certainty that no accident will prevent our reaching our destination. We lay the foundations of a house without assurance that we shall live to inhabit it. We write a book with no guarantee that it will be published and read. We prepare for tomorrow without any certainty that we shall survive the night. Whether or not we profess to have faith in the promises of religion, most of us reveal this deep, vital faith in our daily living.

Such faith is one of the most fundamental characteristics of life. Existing precariously amid natural forces that so readily overwhelm it, thrown into a maelstrom of competition by its own excessive fecundity, life would be impossible without this capacity to act without assurance of success that is the essence of faith. Without this deep organic faith, no seed would germinate, for in the wild the seedling's chances of surviving to flower are pitifully small. No egg would develop into the embryo of an animal that is more likely to perish than to reach maturity. No bird would migrate if it needed assurance of a safe passage. The faith fostered by religion that we shall somehow outlive our bodies is but a more conscious expression of the strong vital faith diffused throughout the living world.

This vital faith appears to penetrate a little farther into the consciousness of the bird who, having already lost perhaps half a dozen nests to predators, hopefully starts to build another. Can it be totally unaware that its undertaking is precarious? Its very evident anxiety when it detects a predatory animal lurking in the vicinity of its nest, its constant suspicious watchfulness appear to reveal its uncertainty about the outcome of the undertaking in which it resolutely perseveres.

As we sit in view of a nest, watching a parent bird patiently incubate, coax sluggish nestlings to swallow their food, warm them, shield them from hot sunshine and beating rain, keep their nest irreproachably clean, all in many cases with the closest cooperation of a mate, searching questions arise in a thoughtful mind. It is difficult to imagine a situation more appropriate for the growth of such psychic and moral attributes as parental love, conjugal devotion, a sense of obligation or duty, perhaps even a conscience. These parents certainly act as though they were motivated by such sentiments. They might serve as a paradigm which human couples, too often less faithful to their obligations, might emulate. But have parent birds or any others developed the psychic and moral attributes which their activities appear to foster?

I cannot tell whether birds feel affection for mates from whom they are inseparable or for offspring that they serve so devotedly, but I am convinced that they have something closely resembling conscience. Just as the state of our health is a measure of the harmony among the organs and functions of our bodies, so the state of our conscience reveals how closely our conduct harmonizes with our guiding rules or principles. When our behavior conforms to the standards that society has instilled in us or we have set for ourselves, our conscience tends to be at ease. When we depart from these guidelines, doing what we believe to be wrong or failing to perform a recognized duty, conscience is more or less acutely troubled. The persistent disagreeable nagging, the sense of shame or inadequacy, of a bad conscience is so distressing that we learn to endure hardships and discomforts, to engage in exhausting toil physical or mental, in order to keep peace with conscience. To preserve a fairly tranquil conscience, as to preserve health, becomes a major endeavor of one who lives wisely and well.

A nesting bird follows an innate pattern of activities, at least as tightly integrated as and hardly less complex than that which a person who has been well raised learns from his parents and teachers. To be unable to conform to this pattern appears to distress the bird, much as a troubled conscience distresses us. It is obviously upset when the presence of an intruding animal, human or other, prevents it from incubating its eggs, brooding, or feeding its nestlings. To preserve the integrity of its pattern of behavior and avoid the feelings of discomfort or unrest which probably accompany its disruption, a bird will frequently suffer privations, as when it gives its young food that might satisfy its own hunger, or in a prolonged rainstorm remains sitting and fasting far longer than its customary intervals, or works so hard for its brood that it loses weight. A bird attending its nest appears to be motivated by something very like a sense of duty.

Recognition that care of dependent young is the context in which such moral qualities as responsibility, sense of duty, faithfulness, and loving devotion were born leads us to ask whether modern welfare states, which relieve parents of a substantial share of their traditional obligations, are not weakening at their source attributes indispensable for a civilized community. When the state makes itself responsible for nourishing the child, with school lunches or otherwise, safeguarding its health, ensuring that it has adequate shelter and medical care, all in addition to the state's older undertaking to educate it, parents whose sense of responsibility may not have been strong in the first place are likely to become increasingly irresponsible. Why should they bother to limit the number of their offspring when, whatever happens, a concerned state will take care of them, perhaps even increase the parents' income with subventions? In such a situation, people deficient in responsibility may breed more rapidly than conscientious, self-respecting citizens, further increasing the burdens that taxpayers must bear to support increasing multitudes of the irresponsible, for children who grow up in an atmosphere of irresponsibility are hardly likely to become conscientious parents and citizens. Moreover, since it is a general evolutionary law that unused organs and faculties tend to atrophy, the innate foundations of moral qualities, no less than their culturally nurtured expressions, will inevitably weaken. And the nemesis of this situation may fall harshly upon the careless parents, for children who have known little of parental devotion, who have looked to the state rather than to their progenitors to safeguard their lives, may fail to develop a strong sense of obligation to their parents, who at an advanced age may find themselves neglected by their descendants and wholly dependent upon the state's coldly impersonal beneficence.

In a quite different context, birds reveal glimmerings of conscience or a sense of right and wrong. It has frequently been observed that when birds with adjacent territories fight at their common boundary, each is dominant on its own side. As the conflict surges back and forth across the unmarked boundary, each in turn can repel the other. Since there is no reason to suppose that a bird's physical power suddenly waxes or wanes with each spatial displacement of a few yards, something similar to a moral inhibition appears to arise, weakening the bird's fighting ability when it is on its rival's land—much as, with us, one capable of valiantly defending his home pilfers his neighbor's fruit furtively, like a coward.

With clear indications of a conscience and sense of duty, birds appear to be well advanced along the road to becoming moral beings, but probably they have not yet arrived. To be fully moral is to be able to ponder questions of right and wrong, to anticipate duties, to feel remorse when we have gone astray. Although a bird's behavior is often what the most exacting morality would demand—indeed, it often puts us to shame—we cannot attribute morality to it without implying certain mental qualities which we are not certain that it has. Let us, then, recognize the bird's approach to a moral state that it has not quite reached by saying that it is protomoral.

This protomorality of the feathered creature is often admirable. Not only does it ensure the efficient performance of the most complex and exacting program of parental care in the whole animal kingdom, apart from man, but in many nonmigratory birds and a few, including geese and swans, that migrate in family groups it also promotes a high degree of monogamous fidelity. Moreover, it brings harmony into a bird's relations with other individuals of its own and different species. A dozen kinds may nest without conflict in the same garden or peaceably share the berries of a generously fruiting tree. In this respect birds have certainly advanced farther than men in tribal societies, whose moral rules applied only to fellow tribesmen and placed no restraints on their treatment of members of neighboring tribes—the state of internal amity and external enmity of which Herbert Spencer wrote.

When we compare the innate protomorality of the more advanced birds with our own more self-conscious morality, which each of us acquires largely from external sources, we may well ask what advantage the latter has over the former. At its best, our morality is more flexible and able to confront a greater range of situations, to deal with the unusual, and to extend itself more widely than the birds' protomorality. Except in a few highly social species, such as the wood-swallows of Australia that Immelmann (1966) studied that may feed ailing or injured companions, the disabled bird falls out of the avian pattern of behavior. Perplexed by this departure from the normal, the bird does not know what to do about it; it may even abuse the handicapped member of its flock. Men are kinder to afflicted members of their society. On the other hand, the greater flexibility of our behavior involves us in perplexities and makes it easier for us to go astray, so that sometimes we wish that we could return to the protomorality of animals.

In an amusing essay, the great entomologist William Morton Wheeler (1920) suggested that we should "ram virtue and health back into the germ-plasm, where they belong," or, in other words, make our morality innate, like the protomorality of birds. If we could do this, we would doubtless enjoy many advantages, but the loss of our more thoughtful, foreseeing morality would be lamentable. Best of all would be the harmonious integration of these two methods of determining conduct. If basic moral virtues such as moderation, honesty, justice, chastity and conjugal faithfulness, parental and social responsibility, generosity, and the like had such strong innate foundations that it would be difficult for us to deviate from them, we would all live more happily in a more harmonious society. At the same time, we might retain our faculty of forming moral judgments and seeking ethical guidance in unusual and difficult situations. Moreover, we would not wish to lose our capacity to cherish ideals that broaden and elevate our morality, making it ever more sensitive to all life's changing circumstances and ever more concerned for the living world that surrounds us. These ideals need not be easy or even possible to realize in the foreseeable future— ideals that we can realize without much effort tend to be too narrow and unworthy of us. Our highest ideals continue to rise above our reach as with sustained effort we climb steadily toward them. To have an innate morality as a firm foundation for moral ideals not inferior to that of birds, who are the most faithful parents and most peaceful neighbors, would be a precious gift.

The birds' strong parental attachment compensates for an inefficient breeding system. They might do much better if they did not hatch their eggs in nests. The hard-shelled egg, so perfect in shape and often so richly colored, is in some ways a beautiful mistake. Although birds do their best to hide their nests or make them inaccessible, these structures are too readily found by a host of hungry predators. When a nest is threatened, the parents may try to repel the enemy, but many are too powerful to be held aloof. Although most parent birds who lose eggs or nestlings live to try again to rear a brood, some are killed on the nest, so that the mortality of incubating or brooding small birds, especially of the sex that passes the night on the nest, tends to be higher than that of the other sex.

It is instructive to compare the reproductive system of birds with that of their mammalian counterparts, the bats, which, much less numerous in species than birds, are possibly equally or more abundant in individuals in tropical woodlands. Many tropical bats have only two pregnancies a year, each time with a single offspring, and the Sac-wing Bat maintains a high population with a single pregnancy, also with only one baby. Probably, since the embryo develops inside the parent, mothers who survive give birth to living young. Two is the number of eggs most frequent in nests of tropical land birds, but in the forests only about half of these eggs escape predation until they hatch.

One wonders why birds are the only class of vertebrates of which every species lays eggs that hatch apart from the parent's body. With the exception of the monotremes, all mammals retain embryos inside their bodies or, as in marsupials, in pouches or otherwise attached to their bodies in the later stages of development. Among the mostly oviparous fish, amphibians, and reptiles are numerous species that hatch their eggs internally or attached to a parent's body, as in certain frogs. To retain eggs inside their bodies until they hatch would also be advantageous to many species of birds in several ways. The embryos would develop at a uniform rather than a fluctuating temperature, as in all those species that do not incubate their eggs continuously. They would be much safer, because the mother bird could fly from approaching predators with them. The parent would enjoy more time to seek food with its mate or flock. Probably, as the date of hatching approaches, the parents would need to prepare a nest in which to feed and brood the tender young; but this course could abbreviate to approximately one-half the period of occupancy of vulnerable nests, with a corresponding reduction of mortality not only of the nests' contents but of the attendant parents. Although female birds probably could not carry as many eggs as are laid by some temperate-zone birds with very large broods, decreased losses would make it unnecessary to produce so many. Northern bats maintain their populations with only one or two young per year.

The great advantage of the avian system is that it promotes close association of the male and female parents and their equal participation in every phase of the reproductive process except the formation of the eggs, something rare in the animal kingdom. This is a value that compensates for the vulnerability of eggs. It is pleasant to know that evolution does not invariably raise efficiency to the highest possible level, which can become repugnant and too often has distressing consequences, as when the too efficient production of offspring leads to widespread starvation and other evils, or great efficiency in agricultural and industrial production overloads the market and brings on economic depression. A biological engineer might improve the efficiency of the avian reproductive system, but at the price of making it less accessible to observation. If it were not contrary to evolutionary theory, we might suppose that birds build nests for their eggs so that we, by watching these nests, might become more intimate and friendly with them.

To sit in thoughtful contemplation of birds devotedly attending their eggs and young affects our attitude toward them and perhaps toward the whole living world. Those who study birds as museum specimens or investigate their diets or physiology may shoot them for specimens or to accumulate data for statistical analyses and graphs, but one who has watched them sympathetically at their nests finds it difficult, if not impossible, to harm them. A creature that is the product of so much self-sacrificing parental care as most birds receive appears to have higher value, to be more worthy of our protection, than an animal that develops from an egg that is deposited and wholly neglected. One becomes indignant at the sight of a snake that cares for nothing swallowing a young bird raised by faithful, perhaps loving, parents. Although the contemplation of certain aspects of nature, such as the prevalence of parasitism and predation, may engender a gloomy outlook, a pair of birds cooperating closely to rear their young is more likely to foster an optimistic world view. For of all classes of terrestrial vertebrate animals, these charming creatures are by far the most numerous in species and probably also in individuals. Evolution, for all its blundering and miscarriages, is capable of splendid achievements.

18. *Versatile Leaves*

Green leaves are organs highly specialized for the single task of making carbohydrates by means of photosynthesis. In this they contrast with other vegetable organs with varied functions. Stems and their branches raise the plant in the air, display its foliage and flowers, conduct water and salts upward and elaborated nutrients downward, store them for future use, and form buds for continued growth. Roots anchor the plant in the ground, absorb water and salts, and often serve as storage organs. Flowers advertise their presence by bright colors and produce scents to attract pollinators, nectar to reward their services, pollen in their anthers, and seeds in their ovaries. But the single task of green leaves, that of synthesizing the food that supports all vegetable and animal life on the continents and islands, is so important that no other has been assigned to them. When, with increasing age, the accumulation of salts in their tissues makes them less efficient at their only task, they fall and are replaced by new leaves. The exceptions to this rule are so rare that they arrest our attention.

A broad expanse of the hillside behind our house is covered by the fern *Dicranopteris pectinata*, which ranges from Mexico to Bolivia and throughout the West Indies. Starting on the cut bank above the road, it has steadily advanced up the pasture on the steep slope until it forms a light green impenetrable thicket, more than head high, where Scarlet-rumped Tanagers and Black-striped Sparrows roost amid the dense verdure and Agoutis seek shelter on the ground beneath them. It has climbed up the scattered small trees in its midst to a height of 20 feet (6 meters). This stubborn tangle is composed wholly of the fern's leaves, or fronds, which spring from rhizomes that creep beneath the ground. Each new frond that, like a long, thin green rod, pushes upward through the mass ends in two branches, which in turn bifurcate repeatedly. On the ultimate divisions, which are nearly always in pairs, short, narrow segments are arranged in the form of a comb.

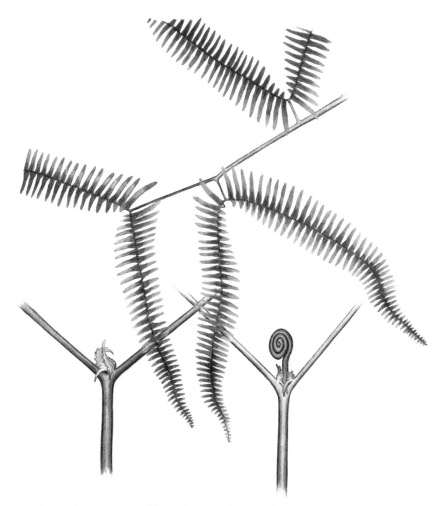

Dicranopteris pectinata, *scrambling fern, with coiled young segment of frond in bifurcation (bottom right)*

On their glaucous undersides some of them bear sporangia in naked clusters.

Between every pair of branches is a bud covered with brown hairs. The well-developed bud at the apex of the stipe, or basal stalk, is shielded by a pair of bracts that resemble small leaflets, as are the buds at the bifurcations of the second and sometimes the third order. Those in the forks of higher order usually remain rudimentary. Soon the principal bud at the end of the stipe elongates to produce another section of frond like the first, and this may be repeated again and again until the leaf becomes 20 feet (6 meters) long with a tough, wiry main axis, the rachis, which the botanically uninstructed would call the fern's stem. With such repeated forking binding all the fronds together, the fern forms a snarl penetrable by no creature more corpulent than a small bird or a snake. Although we often think of ferns as delicate growths of moist, shady places (as some, especially the filmy ferns, undoubtedly are), a few of them are among the most stubborn weeds with which the agriculturist has to contend. Slashing them down at ground

VERSATILE LEAVES 149

level hardly helps, for the underground rhizomes promptly send up new leaves. In our valley, Bracken is the most troublesome, and the dichotomous *Dicranopteris* comes second.

In the rain forest grows another fern with even longer leaves, formerly called *Salpichlaena volubilis*, although it is now considered a scandent *Blechnum*. From a slender rhizome that creeps through the rich loam just below the surface spring clusters of fronds such as one would not be surprised to find in northern woodlands. Rarely more than a yard high, they bear from six to ten pairs of leaflets with an odd leaflet at the apex. Broadly linear, with acute tips and wavy margins, these leaflets are from 6 to 8 inches long by 1 to 1.5 inches wide (15 to 20 by 2.5 to 4 centimeters). Only on an exceptionally vigorous frond, which may become 4 feet (1.2 meters) long and bear eleven pairs of leaflets, are the basal leaflets compound instead of simple, suggesting a remarkable change to come.

Abruptly, with no transitional forms, the rhizome sends up a slender, green, wirelike flagellum, which may be quite smooth for the first 3 to 5 feet (.9 to 1.5 meters), corresponding to the stipe of the frond, beyond which it bears tiny rudiments of lateral leaflets which will not develop unless the flagellum finds a support. It ends in a scroll or fiddle head like that of any growing fern frond. I found one of these thin green cords that trailed through the low fronds of neighboring ferns to the length of 11 feet 4 inches (3.45 meters), of which the first 58 inches (1.47 meters) corresponded to the naked

Blechnum volubilis, climbing fern, with terrestrial fronds, a flagellum (left), and frond twined around trunk

stipe of a typical fern frond. When the groping flagellum finds a support, such as the trunk of a palm or a small dicotyledonous tree, it twines around it, ascending from left to right, just like a typical vine. Now this scandent growth, which is an elongated, twice-pinnate frond, bears paired divisions which resemble the smaller of the fronds on the ground, with the difference that each pinnate primary division corresponds to a simple leaflet of the basal fronds. Among them, at certain seasons, are fertile divisions which resemble the others but have much narrower leaflets, each with a continuous row of sporangia covered by a brown, incurved, protective membrane, the indusium, along each side of the midrib.

These fronds spiral up trunks to a height of 40 feet (12 meters). Probably if unwound they would be little short of 50 feet (15 meters) long, but I have not succeeded in separating one from the tangle that a number of them form on a favorable tree. It would be difficult to find longer leaves anywhere. By endowing some of its fronds with the capacity for indeterminate growth, the fern rises above the dimly lighted undergrowth of the tall forest to a region of greater light, as neighboring tree ferns do by means of tall, stout, erect stems. Although this climbing *Blechnum* never forms such dense tangles in the shady depth of the forest as the *Dicranopteris* does in full sunlight, it is often difficult to pass through a maze of its wiry fronds laced through the undergrowth.

I have found only one tree whose leaves continue to grow as these fern fronds do. In wet, epiphyte-laden forests of the Costa Rican mountains, between 5,000 and 6,000 feet (1,525 and 1,830 meters) above sea level, grows a tree of medium height that attracts attention by the great compactness of its narrow crown. Even when it stands in a clearing, where many trees spread their branches more widely, *Guarea rhopalocarpa*, of the Meliaceae or mahogany family, retains its peculiar growth form. Examination of its foliage reveals the reason for this. Its newly expanded leaves are pinnately compound, like those of the ash and the walnut, with usually two or three pairs of oblong, acuminate, opposite leaflets. Instead of the single terminal leaflet which many such leaves have, the axis of the leaf ends in a bud, in which one can distinguish the rudiments of several pairs of leaflets folded together upward with their margins rolled inward and beyond them a growing point, as at the apex of a typical stem or branch. As is usual in the tropics, the bud lacks the scales that protect the leaf buds of woody plants in lands where winter is severe, but it is densely covered with short hairs, just as are the buds at the ends of the branches of the same tree.

In February, when the terminal buds of the branches elongate to produce new leafy shoots, the buds at the tips of the leaves also become active. Each grows into a new length of rachis bearing two or three pairs of new leaflets or sometimes only one pair. Until nearly full grown, these leaflets are deflexed below the rachis with their lower sides facing one another. For some time after they stop growing, these new leaflets may be distinguished from the persisting older ones by their much lighter, fresher green.

The growth of the leaf is renewed in this fashion from year to year. Whether this occurs annually or sometimes at shorter intervals, I did not learn. Although the axis of the leaf becomes long, it always remains slender and withelike, drooping under its own weight and that of the leaflets it bears. The longest leaf that I could reach measured 50 inches (127 centimeters) from its insertion on the branch to the bud at its tip, but it was only a quarter inch (6 millimeters) thick at the base. It had produced, as indicated by living leaflets and scars where older ones had fallen, a total of twenty-two pairs, including two newly expanded. Allowing for the production of two or three pairs of leaflets each year, its age would have been from seven to eleven years. Possibly it was even older, for sometimes only a single pair of leaflets is added when a bud elongates. In any event, it was exceptionally long-lived for a leaf. The leaflets do not last so long, for I found no more than five or six pairs on a leaf.

The basal part of the rachis, from which the leaflets have fallen, remains quite naked because the leaflets lacked buds in their axils, as one would expect to find if these were simple leaves on a slender branch instead of leaflets on the rachis of a compound leaf. In other respects, the axis closely resembles a slender branch, in both external appearance and internal structure. It becomes covered with corky bark; within, it has a cambium that forms a cylinder of wood with growth rings. Although we associate cambium with stems and branches rather than with leaves, I found it present, often forming a closed cylinder of woody cells, in the petioles and rachises of a number of trees with compound leaves, including the Large-leaved Jacaranda and the Flame-of-the-Forest tree, and even in the petioles of such large simple leaves as those of the Cecropia and Castor-oil Plant. But because these other leaves never renew their growth in the manner of the *Guarea*, I could detect no rings in their wood. They also lacked cork, which is far from common on petioles.

Although leaves that continue to grow are not common, they are found, in addition to numerous species of *Guarea*, on a number of unrelated plants in a wide variety of habitats. About the only thing that the gymnosperm *Welwitschia mirabilis* of the severely arid deserts of southwestern Africa has in common with *Guarea rhopalocarpa* of humid mountain forests is the continued growth of its leaves. From a tough, tapering root, this extremely hardy plant produces a stem or trunk which may become 5 feet (1.5 meters) thick but rises little above the ground. In addition to its short-lived cotyledons, *Welwitschia* bears only a single pair of opposite leaves, which spring from a deep horizontal groove at the top of the taproot and continue to elongate from their protected bases for the whole of the plant's life of hundreds or possibly thousands of years. At first shaped like broad ribbons with parallel veins, the leaves grow to a length of 10 or 12 feet (3 or 4 meters) and split into numerous strips where they lie on the barren, sun-baked ground.

Although they do not continue to elongate, large compound leaves may substitute for branches. The twice-compound leaves of young trees of the Large-leaved Jacaranda are often 6 or 7 feet long by 1 yard broad (1.8 or 2 by .9 meters), bear over a thousand leaflets of medium size, and have the photosynthetic capacity of a whole branch. Their axis is so hard and stout that it may be used as a walking stick or a child's play horse. Until the tree is 40 to 100 feet (12 to 30 meters) high, it bears no branches but only these broad leaves that resemble those of tree ferns. Their task of photosynthesis done, the jacaranda's leaves fall, leaving big, prominent scars on the trunk. Thus, they give no lasting hold, as more permanent branches would do, to the profusion of vines and scrambling shrubs that in the second-growth thickets frequently overwhelm young trees that try to rise above the riotous growth amid which jacaranda trees frequently start life. Similarly, the huge compound fronds of palms and the broad filigree fronds of tree ferns are leaves that substitute for branches on tall, branchless trunks.

Leaves may serve as trunks. The massive column, often 6 inches (15 centimeters) or more thick, that bears the banana plant's spreading crown of huge leaves 15 feet (4.5 meters) in the air is readily mistaken for a woody trunk. Nevertheless, a single stroke of a sharp machete can sever this false trunk, as it could never do with a woody trunk of equal thickness. The severed banana trunk reveals an attractive pattern of crescentic leaf-bases, each tightly overlapping those within. The trunk is so easily cut because it contains no wood, and the leaf-sheaths that compose it are full of air spaces—the stately banana plant is but an overgrown herb.

One can dismember a banana trunk by pulling down the leaves from outside inward. Each consists of a blade up to about 13 feet long by 1 yard broad (4 by .9 meters) that was at first a single broad expanse but is often split into ribbons or wider sheets by wind or rain. This lamina is connected by a stout petiole to a long, narrow, incurved sheathing part that helps form the trunk. The leaf's total length may approach 30 feet (9 meters). As more and more leaves are pulled away, the trunk shrinks in diameter until the removal of the last exposes the true stem, thin, smooth, and white. This glossy rod bears a few smaller leaves as it pushes upward through the center of the column of overlapping leaf-sheaths to emerge at the top, bend over, and develop a huge flower bud that yields a bunch of fruit so heavy that the weak stem could never support it without the aid of the leaves which tightly embrace it over most of its length.

Leaves sometimes serve as tendrils. The simple leaves of the Climbing Lily have long, slender tips that coil around convenient supports. The compound leaves of Sweet Peas and of their relative in the bean family, the huge, aggressive tropical liana *Entada gigas*, terminate in branched tendrils. The climbing species of Virgin's Bower, *Clematis*, depend upon their long petioles to curl around supporting twigs and uphold them. All these leaves that act as tendrils have green expanses for photosynthesis.

The fleshy leaves of the stonecrop family and other plants of arid habitats store much water in their tissues. The tiny, thin leaves of liverworts, so dependent upon rain and dew to keep them moist, retard desiccation in another manner by coiling into bladderlike shapes that retain water. On branches of the Calabash trees around our house and on rocks in the pasture, the liverwort *Frullania* forms thick, brown cushions. Beneath each of its tiny dorsal leaves is a minute flask, in which a rotifer nearly always dwells. Much of the time the diminutive wheel animalcule remains dormant, for these liverworts thrive where they are exposed to bright, drying sunshine; but if one mounts a fragment of a plant in water and examines it through a microscope, he may watch one of these creatures stick its head through the orifice of each flask and set its twin sets of cilia into wheellike motion, creating currents that may waft microscopic organisms to its mouth. Many of the leaf cells of the bog moss *Sphagnum* lose their contents and develop holes in their outer walls so that they suck up water by capillarity. Even on a sunny day, water may be squeezed from a handful of this pale green moss as from a sponge.

Some leaves supplement flowers and seeds in the reproduction of plants, or they may even play the leading role. The ability of leaves to produce new plants is more widespread in ferns than in flowering plants and among the latter more frequent in dicotyledons than in monocotyledons. In temperate North America, the Bladder Fern bears, beneath its rachis and pinnae, tiny bulblets which readily grow into new plants. On shady outcrops of rock, the long, slender tip of the Walking Leaf roots where it touches soil, producing another plant which likewise reproduces in this fashion until the fern appears to walk over the ledge. In the rich black loam between rocks in tall second-growth woods beside the Río Peñas Blancas grows a much more vigorous "walking fern." Each of its larger fronds ends in a ribbonlike projection that may become a yard long and bear three buds which strike root and grow into new plants where they touch the ground. By means of these long ribbons of green tissue, *Leptochilus cladorrhizans* advances by giant strides through the deep shade of the underwood. With it grows a species of *Dryopteris* with a single bud near the tip of each more vigorous frond.

In the shade of orange trees in the pasture grows a charming star fern which bears, on slender, hairy petioles that rise directly from the ground, leaves shaped like stars with five broad, unequal rays. Its sporangia, devoid of protecting indusia, occur in thin lines that form a network over the whole underside of each horizontally held fertile leaf. In the deep angle at the base of a terminal lobe is a bud that can sprout into a new plant. This may be on either the right or the left side but not on both. However, on old fronds that lie upon moist ground, additional buds give rise to plantlets at various points on the margin. These ferns that grow near me are only a few of the many kinds with buds on their leaves.

Many liverworts bear on their leaves multicellular gemmules that readily drop off and sprout into independent plants. Some of these green, gemmule-bearing leaves, notably in the tropical genus *Colura*, are coiled into water-holding sacs so that they have three functions: photosynthesis, water storage, and reproduction—a rare combination of aptitudes that we find also in a few flowering plants with fleshy leaves.

Of flowering plants that reproduce from their foliage, probably the most familiar is the life plant, *Bryophyllum calycinum* of the stonecrop family, often grown in northern greenhouses and homes and a widespread "weed" of tropical lands. The thick green leaf has a crenate or notched margin, and in each notch is a rudiment that can grow into a new plant. These plantlets rarely develop on leaves still attached to vigorous plants. Only one or a few shoots spring from a whole detached leaf; but if the *Bryophyllum* leaf is cut into as many pieces as it has notches and kept moist, every one can produce a plantlet. This suggests that the intact plant somehow inhibits the growth of the rudiments on its foliage and on a detached leaf the plantlets that grow soonest inhibit the development of others, which makes this species a fascinating subject for experimentation. The formation of buds or plantlets on leaves still attached to the parent plant is widespread in the stonecrop family of arid lands, but I do not know a single species of flowering plant in the luxuriant native vegetation around me that reproduces in this manner.

As is well known to horticulturists, many species of begonias and the Bowstring Hemp, *Sansevieria* of the lily family, among many other plants with fleshy leaves, are readily propagated from detached leaves or parts of them. Of the common Bowstring Hemp there exist two varieties, one with light green leaves crossed from edge to edge by wavy bands of darker green; the other, less common, with broad yellowish margins devoid of chlorophyll. It is a curious fact that the all-green variety is readily propagated from its leaves, but the pale-bordered variety can be reproduced only from the creeping rhizomes.

Finally, we may notice, without expanding upon so large a subject, that the leaves of a considerable number of plants have become traps to catch and digest insects and other small creatures. The leaves of sundews are covered with stalked glands that secrete a viscid gum which holds the unfortunate insect attracted by the glistening drops while neighboring glands bend over to grasp it more firmly and digest it. The small leaves of Venus's Flytrap, hinged in the middle and fringed with long, slender teeth, snap inward to capture the incautious creature that touches sensitive bristles on their upper surfaces. The deep tubes of pitcher plants are traps well fitted to prevent the escape of small invertebrates that fall into them. When an animalcule swims against the trapdoor of a set bladder of the bladderwort, it opens, permitting the tiny sac to expand suddenly and suck the creature in, as by a pipette. The only insectivorous—or, more properly, carnivorous—plant that I have found in this region is the bladderwort *Utricularia endresii*, which grows on mossy trunks in the foothills over which I look. Its basal leaves with tiny bladders are embedded in the moss, above which slender stems bear large lavender or purple flowers that are readily mistaken for orchids.

Despite the varied functions that some leaves perform, it remains true that those of the vast majority of vegetable species do nothing that might interfere with their important primary task of capturing the energy in sunlight to synthesize the food that nourishes not only the plants but all terrestrial animals.

19. *The Convivial Ascetic*

At first sight, the title of this chapter appears to involve mutually contradictory terms, like "black whiteness" or "the son of a childless woman." Everyone knows that a convivial person is given to eating and drinking in jovial fellowship; we commonly think of him as seated or reclining at the festive board. The ascetic, on the contrary, holds himself aloof from such gatherings. Although he may dwell with others who share his austere habits, he is often a hermit who munches in contemplative solitude his coarse and frugal meal. The festive board where wine flows freely is anathema to him. How could an ascetic be convivial?

Such is the common conception, but if we analyze the situation a little more deeply we shall see that it is wrong. At least, the usual distinction between the convivial person and the ascetic fails to do justice to the literal meaning of the word convivial—*con vivere*, to live with others. Let us in imagination seat ourselves, as uninvited and unseen guests, at the groaning board; and while the invited guests lift the wine cup in repeated toasts, stuff themselves with an excess of rich foods, and pass the merry quip and shake with laughter, let us reflect earnestly upon what we behold.

In most countries, and throughout the greater part of history, our thought would trace this superabundance of choice comestibles back to the toil of oppressed serfs or driven slaves, living in misery and squalor, deprived of adequate nourishment, bereft of freedom, beauty, and hope, so that their masters might enjoy enervating luxury and ease. Yet many of these downtrodden toilers are in natural endowment nowise inferior to their lords; given the same advantages, they would in numerous instances surpass them in wisdom or wit. But, dull and uncouth as the inevitable result of the harsh conditions of their lives, they are naturally excluded from the brilliant company amidst which we sit. The convivial fellowship does not extend to them; it is confined to an elite, too often callously indifferent to the well-being of the very workers upon whose toil it subsists.

But, it will be objected, the situation we have just contemplated is archaic; in the more advanced countries of modern times, where agriculture no less than industry is largely mechanized, everyone enjoys an abundance without the forced labor of slaves or of serfs bound to the land. While there has been much improvement in this respect, it is still true that the world contains many human mouths that might benefit by the food which harms those who eat too lavishly. Yet, even with the assurance that the meat and drink that burden the table at which we sit were not produced by undernourished laborers, we are not quite happy about them. Our mental vision follows the flesh on the plates before us back to the reeking slaughterhouse, where inoffensive animals in the prime of life were cruelly butchered, perhaps after a long, harassing journey in an overpacked van, after much abuse and mutilation on the farms where they were raised. Even the ice cream, so innocent in appearance, calls up visions of calves who never enjoyed a taste of their mother's milk, who were perhaps slaughtered at a tender age to increase the dairyman's profits.

The cigarettes, the coffee, and the cocktails appear to escape this objection. We cannot by any stretch of the imagination detect a drop of blood upon them. Yet tobacco, coffee, and the fruits or grains of which alcoholic drinks are made are grown on fertile land that is thereby excluded from producing food which some people, perhaps the very laborers on this land, could well use. And although the area needed to supply these things for a single person may not be great, the aggregate consumption by teeming modern populations represents a huge drain on Earth's bounty, which even without this additional burden is strained almost beyond its limits by humanity's multitudinous demands. It is clear that the greater man's consumption of unnecessary luxuries becomes, the more he reduces the areas of Earth which remain to support the natural vegetation and the free animals of all kinds that dwell in it. These also seem to deserve a portion of this planet, to whose beauty and interest they contribute greatly. Thus, despite the emancipation of slaves and helots, the feast of which we are the unseen spectators has not achieved true conviviality in the literal meaning of the word. This sort of entertainment is not living with other creatures to the best of our ability. Our merrymakers are still members of a little closed society, thoughtless of the wider fellowship of living beings.

Let us now turn to the ascetic and see how the situation stands with him. He sits alone, eating his frugal meal of rice or bread, garden vegetables and fruits, washed down with pure water. Not for him the rich, surfeiting viands, the wine that loosens tongues and weakens self-control. How unsocial, how brutish his solitary meal in contemplative silence! But this is to see only with the corporeal eye while that of the spirit remains tightly shut. The ascetic is not alone; he shares his repast with unseen companions. Because of his abstemiousness, many a creature of the most varied kinds, which if he followed the thoughtless, luxury-seeking existence of the multitude would have been either directly or indirectly destroyed, is now enjoying life. Other people, too, have more because he takes less. Whether near or far, these beneficiaries of his frugality are his commensals, the sharers of his meal. He lives with them rather than at their expense. He, not the feaster at the overladen table, is the true convivialist in the literal meaning of the word.

Doubtless, to many it appears a perverse use of intelligence to permit our wandering thoughts to destroy our spontaneous enjoyment of a situation which the natural man regards as a source of great pleasure, while it exalts a mode of life that to him is anything but attractive. But this reversal of naïve appraisals is the inevitable result or expression of spirituality, as I understand the term. Placed at the festive board that we earlier considered, the intellectual man, the analytic thinker, may see clearly enough the relationship between the superabundance that he enjoys and the deprivations and sufferings of certain other beings, human and nonhuman. If he is merely intellectual, these thoughts will not in the least diminish his enjoyment of present sensations. The spiritual man, however, cannot find pleasure in sensations or experiences that he perceives to be purchased at the price of others' pain. Insofar as the creatures who suffer to provide gratifications for him are not immediately present, he must possess analytic intelligence to trace the connection between his actual sensations and the unseen sufferings of others.

Thus spirituality, in any high degree, appears to consist of intelligence plus something else. One of its aspects is the capacity to have our enjoyments or our sufferings heightened or diminished by insight into the wider relationships and more remote consequences of our activities. Accordingly, the spiritual person differs from the sensual person by having both greater intelligence and more responsive affections and from the merely intellectual person by his greater emotional sensitivity. He adds zest to his meager repast by picturing to himself the creatures that benefit from his frugality; he abhors the wanton banquet because he cannot close his spirit's eye to the suffering and destruction for which it is responsible.

It appears, then, that asceticism, not in its harsher aspects but at least to the extent of thoughtful moderation, is the inevitable reaction of the spiritual person who contemplates his situation as an animal with rather large and complex needs in a world crowded with sentient life. His frugality is the most perfect expression of his awareness of and solicitude for the multiform life around him. The ascetic does not scorn happiness; like everybody else, he is compelled by the constitution of his own mind to seek the greatest happiness, the highest ultimate good, as he sees it. All the great ascetic systems, such as Stoicism, Jainism, and monastic Buddhism, have carefully planned procedures to ensure perfect and unshatterable happiness to those who consistently practice them. The founders of these disciplines saw that unrestrained indulgence in sensual pleasures is not the road to happiness, for such gratifications are all too commonly procured at the price of much suffering by other creatures and even a balance of pain by one who pursues them.

Asceticism, of the sort that strives for true conviviality, need not and should not extend to things of the spirit. Born of spirituality, it becomes untrue to its source when it cramps or depresses the spirit's life. All the felicity that can be won by the contemplation or pursuit of truth, beauty, and goodness and from the cultivation of friendship seems wholly compatible with asceticism in things of the flesh. Indeed, frugality in food and drink, the avoidance of excessive luxury, is the regimen that best prepares the spirit for a satisfying life in its own sphere.

Our crassly materialistic modern civilization commonly underestimates the mind's capacity to create its own felicity with few material supports. Even in early childhood it demonstrates this capacity to a remarkable degree. When children straddle a stick and imagine they are riding a spirited horse, when they set three chairs in a row and fancy they are taking a long journey on a railroad train, must we not concede that the material component of their enjoyment bears about the same proportion to the mental component as the mass and complexity of a stick to those of a horse, or those of three chairs to a railroad train? Apparently, this spontaneous tendency of the spirit to lead its own life more or less independent of the physical milieu has, in an evolutionary way, been repressed by the necessity to take a more realistic view of things in order to survive, and society for its own ends leads the child out of the realm of fancy into the harsh realm of economics. Yet the sight of children at imaginative play should serve to remind us in what region happiness is to be found.

There is no reason why the convivial ascetic should not enjoy to the full the tropical splendor that he tries to preserve. Doubtless there are also social gaieties compatible with the kind of asceticism that we have been discussing, because they are innocent in the sense that they do not rest upon the exploitation and suffering of other beings. But it is not easy to find such innocent social diversions except within the context of an innocent society. Or if perfect innocence is incompatible with survival in a world like ours, we should at least demand a society imbued with compassion, which makes innocence an ideal that it strenuously strives to realize. In a civilization whose very festivities frequently reek of exploitation and the slaughterhouse, innocent social diversions are somewhat difficult to find. In such an ambience, the ascetic who would be truly convivial must, paradoxically, pass much of his life in solitude.

20. *A South American Marsh Bird*

As I approached a small pond in northern Venezuela, a volley of grating, rasping, and churring notes, alarming in its sudden loudness, greeted me. Coming closer, I found the author of this vocal outburst clinging to the nodding tip of a stalk that rose above the dense marginal stand of grasses and regarding me with piercing yellow eyes. About 8.5 inches (22 centimeters) long, he was a bird of striking aspect, elegantly attired. His whole head and hindneck were black; his wings and back deep brown, which brightened to rufous or chestnut on his rump and upper tail coverts. His strongly graduated tail was black, with broad white tips on all but the central pair of feathers. On each wing was a large patch of white, visible only when he flew. On each side of his neck, below the black, was an area of deep yellow bare skin. His under plumage was buff, with narrow, distant dark bars on his flanks. His bright golden eyes, gleaming intensely in his glossy black head, gave the impression that nothing escaped his penetrating gaze.

His thin black bill, long tail, and slender body, which fitted him for slipping through the dense marsh vegetation amid which he dwelt, gave him an aspect of streamlined grace. He was my first Donacobius, a bird of uncertain affinity widespread in swamps, marshes, and river bottoms from eastern Panama, Colombia, and Venezuela to Bolivia, Paraguay, and northeastern Argentina and from the lowlands up to about 2,500 feet (750 meters) above sea level.

Since, to my knowledge, nobody had carefully studied this unique bird so widely distributed through South America, I was eager to learn the habits of the few pairs that lived at La Araguata, the hacienda among the hills south of Valencia in the state of Carabobo where for four months I studied birds. I found them chiefly in pairs, in which the sexes looked quite alike, sometimes with dependent young. They subsisted largely if not wholly on insects, spiders, and other small invertebrates that they plucked from the marsh vegetation or picked

Black-capped Donacobius

from the surface of the water between the crowded stems, where much of the time they lurked unseen.

In Venezuela, this bird is known as Paraulata de Agua—the Water Paraulata—to distinguish it from an array of large songbirds, from thrushes and mockingbirds to saltators, also called *paraulata* with appropriate modifiers. In bird guides, it is called Black-capped Mockingthrush. Since there is nothing thrushlike in its appearance, habits, or habitat, I expected that at least it would be a mimic, like other mockingbirds. When, after several months of intimate acquaintance, I had failed to hear the bird imitate anything, I concluded that mockingthrush is an inappropriate name and called it simply Donacobius, its scientific designation, which means reed dweller.

The Donacobius has a number of loud, clear, ringing notes, each of which may be repeated rapidly a number of times, forming an utterance more noteworthy for the power and excitement that it suggests than for its beauty. *Cheeo cheeo cheeo cheeo cheeo*, and *chu chu chu chu chu*, and *whoi-it whoi-it whoi-it whoi-it* are versions that I recorded. Once I heard a male deliver a clear, undulatory song with his lower mandible rapidly vibrating. These notes were uttered with the mouth widely open, revealing its black interior. I believe that only the male sings in this fashion. Such loud, vehement singing seems too exhausting to be long continued. The few Donacobiuses that I knew never indulged in prolonged, freely flowing song such as one hears from other members of the mockingbird family and many kinds of thrushes and finches; they sang chiefly when stirred by some excitement such as a threat to their territory. Possibly in extensive marshlands where these birds are more abundant, they stimulate each other to greater vocal activity.

These birds were surprisingly quick to detect my approach to the marsh where they lurked unseen. Often my first intimation of their presence was a startling outburst of harsh, protesting notes which, far more than their song, suggested their affinity to mockingbirds and Gray Catbirds.

In early May, two pairs of Donacobiuses proclaimed their presence in and around a dense stand of broad-leaved grasses, higher than my head, that filled a moist hollow between pastures. This area of marsh grass, about 100 yards long by 12 yards in greatest width (91 by 11 meters), was well shaded by *Erythrina* and other trees. From time to time, especially in the early morning and in the evening, a volley of loud, clear notes or else of harsh, rasping sounds revealed that these two pairs had not yet settled their territorial claims.

The members of a pair followed each other closely and often perched only a few inches apart, at times almost in contact with each other. While resting side by side or perhaps clinging one above the other on an upright stem, they engaged in a curious mutual display such as I have seen in no other bird. Each partner spread its long tail until the pattern it presented was a wide, dark, central band broadly bordered with white. Simultaneously, the two birds wagged their fanned-out tails rhythmically from side to side through a wide arc, and while so engaged they opened their black bills to emit contrasting notes. The male uttered a loud, liquid, ringing *whoi-it whoi-it whoi-it* or sometimes a higher note while his mate accompanied him with a sizzling or grating sound. One morning when the territorial dispute was at its height, this performance was repeated at short intervals by both pairs. When the display was at highest intensity, the bird's back was humped up, its tail depressed, its head lowered, and its throat grotesquely distended, doubtless to provide resonance for the loud notes. At lower intensity, the birds wagged their tails and called with their bodies in a more upright posture. Twice I saw a pair display in this fashion while one member held a loose mass of fibrous material in its bill. Occasionally, after a slight altercation, the four birds perched in pairs a few yards apart and all displayed simultaneously. I saw no fights.

I searched fruitlessly for a nest in this tract of high, shaded grass for which the two couples so zealously contended. Not long afterward, however, I found a pair building beside the pond a few hundred yards away and followed the complete cycle of their nesting. Secure in the possession of their territory, these birds did not so often engage in the mutual display.

The nest was situated beside the small pond, about 100 feet (30 meters) wide, made by damming a rivulet where it emerged from a narrow ravine. On both sides of the pond rose steep slopes covered with dense scrubby growth and light secondary woods; below the dam, open pastures and small marshy areas extended to a larger stream. Although at the head of the pond was a fairly extensive stand of densely growing, broad-leaved marsh grass where the Donacobiuses often foraged, they chose to build their nest in a narrow strip of the same kind of grass only a few yards wide along the edge of the pond beside a cowpath Here, fortunately, it could be reached without wading into the open water, which was inhabited by several small and middle-size alligators. Although a variety of flycatchers and other small birds frequented the pond and nested around its margins, the only aquatic bird to be found there day after day was a lone Purple Gallinule.

On May 21, when I discovered this nest by seeing a Donacobius carry something to it, it was a loose, formless mass of fibrous vegetable materials and bits of dead grass blades attached to the broad-leaved grass about 2 feet (61 centimeters) above the watery mud. In the following days, I spent many hours watching for the birds to build, but observation was difficult because the nest was screened by the surrounding grass and I hesitated to expose it lest the birds desert at this early stage. Both members of the pair took contributions to the nest; but one, evidently the female, seemed to do much more than the other, although even she hardly exerted herself strenuously. The male preened interminably in a shrub growing on the neighboring shore. Sometimes, seeing his mate approach the nest with material, he hurried toward it with empty bill, but I could not see what he did there. When a mango fell from a tree on the shore, striking the water with a loud splash, he called with strong, clear, ringing notes.

A few yards from the nest of these Donacobiuses, a Vermilion-crowned Flycatcher was also starting a nest 6 feet (1.8 meters) up in a shrubby *Jussiaea* with yellow flowers. While she and her partner were absent, a Donacobius went to the loose mass of material, plucked a liberal billful from it, and carried it to its own nest. Apparently stimulated by this action, the other Donacobius did the same. A little later, the pair of flycatchers twittered together close beside their newly begun nest. This stirred a Donacobius to attempt another depredation, but it and likewise its mate, who approached with similar intentions, were held aloof by the darts of the flycatchers. After the flycatchers flew away, however, one of the Donacobiuses, who had waited in the lower part of the shrub, ascended to the nest and gathered a large billful of white, cottony material mixed with fragments of fine grass inflorescences. On the way to the nest, the thief opened its bill and dropped all the stolen goods, probably because they were not suitable for its own construction. In all, the Donacobiuses tore four large billfuls from the flycatchers' accumulation while I watched.

The nest amid the marsh grass progressed slowly. I saw no more than seven billfuls of material carried by the Donacobiuses in an hour, and two of these were dropped before they reached the nest, possibly because the birds noticed me spying on them from across the pond. However, I may have missed other trips to the nest because the builders approached through the grass with greater secrecy. The last time that I saw a bird take material to the structure was on May 30, so that building continued for at least ten days and probably a few more.

The nest had now become a deep, bulky cup composed chiefly of fibrous materials of vegetable origin and narrow strips of grass blades or the like. This material was looped around three upright dead grass stalks and, more loosely, around one living grass blade. The lining consisted mainly of narrow strips of grass blades and similar materials. On one side of the strongly incurved rim was a large scrap of lizard skin and on the other a crumpled piece of waxed paper. The asymmetrical nest was 4 inches high and 6.5 by 4.5 inches (10 by 16.5 by 11.5 centimeters) in overall diameter. The cavity was 2.5 inches deep by 3.25 by 3 inches (6.5 by 8.5 by 7.5 centimeters) in diameter at the mouth. This nest was attached to the grass stalks on one side only, with the result that before the nestlings were feathered it tilted so strongly that I propped it up with a forked stick.

On July 18 I found another pair of Donacobiuses building in a quite different situation. Their site was in the midst of an extensive stand of head-high Guinea grass on a low-lying, level riverside *vega*, over 100 feet (30 meters) from the stream. Although by this date hard rains had been falling for two months, the ground below and around the nest was neither inundated nor muddy. This nest was supported at a height of 44 inches (112 centimeters) between a green shoot of a shrub of the composite family and a stem of the Guinea grass, about both of which its materials were wrapped. By July 24, when it appeared to be finished, it was 9 inches high by 6 inches (23 by 15 centimeters) in diameter at the middle where it was thickest. At the top it was 4.5 inches (11.5 centimeters) in diameter. This extraordinary structure, over twice as high as the first, was composed of fibrous rootlets, slender dead vines, tendrils, broad strips of monocotyledonous leaves, strips of bark, and the dark fungal rhizomorphs known as vegetable horsehair. The cup was lined with fairly broad grass blades, with a few fine rootlets and green grass inflorescences on the bottom. The nest contained at least two scraps of reptile skin, and a weft of cocoon silk had been placed on the rim. No egg had been laid by the time of my departure on July 24.

Although the nest beside the pond appeared finished by the end of May, I found no egg in it until June 6, so that well over two weeks elapsed between the start of building and the beginning of laying. One egg was deposited daily until the set of three was completed on June 8. These eggs were unlike any others that I have seen. At the first glimpse, they appeared to be uniform, light reddish brown. Closer scrutiny revealed that they were mottled rather than uniformly colored, but so densely as to cover the whole surface with slightly varying shades of reddish brown. The pigmentation was somewhat heavier at the thicker ends of the eggs. If they could be said to have a ground color, it was a slightly paler rufous brown. When the glossy shells were wet from the frequent showers of this season, they were exceptionally beautiful.

Observation of incubation, as of building, was carried on under a handicap. I was exceedingly anxious not to lose the only nest of the Donacobius that I had, after much searching, so far found. So as not to jeopardize it by exposure, I refrained from cutting any of the surrounding grass, merely parting the blades slightly to permit a view of the nest from across the pond; but, moved by the breeze or by growth, the herbage would gradually close together and hide the nest while I watched. When bright sunshine penetrated the grass and the bird sat with her neck stretched up, her golden eye was easy to detect through my binocular. At other times, I was often uncertain whether she was present.

After I had watched for a while, I became aware that the right central tail feather of one member of the pair, apparently the female, was only half grown, so that it served for identification. Thereafter, I saw only this bird cover the eggs. Sometimes I failed to notice her secretive approach or departure through the surrounding grass; but I managed, on two mornings, to time nine full sessions, which ranged from 11 to 43 minutes and averaged 22.4 minutes, and an equal number of recesses, ranging from 7 to 25 minutes and averaging 14.6 minutes.

While the female incubated, her mate spent much time resting in a shrub on the shore a few yards from the nest, preening interminably. Often the lightly barred buffy feathers of his flanks were prettily spread over his dark folded wings. Sometimes he sunbathed, leaning away from the rising sun with his plumage puffed out, head depressed, bill gaping, sunward wing raised almost vertically above his back, displaying the broad white band at the base of the primary feathers that was usually concealed except when he flew. Although at long intervals, as when he heard another Donacobius in the distance, he sang loudly, his habitual silence contrasted with the songfulness of the Bare-eyed Thrush, Streaked Saltator, Buff-throated Saltator, and Rufous-browed Peppershrike, all singing profusely in the sunshine on the surrounding slopes. Once the male Donacobius seemed to feed his incubating partner; but I could not see what, if anything, he held in his bill as he approached the nest through the dense grass.

One morning in mid June, the male and female often joined in the tail-wagging display already described during the recesses of the latter. In the intervals of displaying and preening her own plumage, the female again and again gently billed the feathers of her mate's neck. Often, as she returned to her eggs, he escorted her to near the nest. Only three days before the eggs hatched, one of the parents took to the nest a limp fragment of waxed paper in which my breakfast sandwiches had been wrapped.

One egg had hatched by the evening of June 24, and by the following noon all three had hatched, seventeen days after the last was laid. The empty shells were soon removed by the parents. The nestlings were devoid of down. At first pink with a dusky tinge, their naked skins became darker in the following days. The interior of their mouths was dark flesh color, and after a few days I noticed that the tongue was marked with three conspicuous black spots, one in front and two side by side behind. Such a pattern is rare among the birds of tropical America.

When these nestlings were five days old, their eyes were opening. A day later, the sheaths of their body feathers were pushing through the skin, and those of the wing plumes were already becoming long. When eight days old, the nestlings wore long pinfeathers. The interior of their mouths had become purplish flesh color, still with the three black spots on the tongue. At ten days, their body plumage started slowly to expand, covering their heads and backs by the time they were two weeks old.

Both parents fed the nestlings, bringing them a variety of small insects and spiders held conspicuously in the tips of their black bills, one at a time. Their droppings were either swallowed or carried away in the parents' bills. As far as I could tell, only the female brooded. Whenever I approached the nest, before and after the eggs hatched, the parents protested loudly with harsh rasping and churring notes, often advancing close to me. After I left, they would promptly go to the nest.

Late in the afternoon of July 11, a young Donacobius was sitting on the edge of the nest. While I watched a short distance away, it hopped off through the crowded upright blades of the marsh grass and vanished, leaving its two siblings resting in the nest. By the middle of the following forenoon, another nestling was perching beside the nest and fled as I approached. Two hours later, the last of the brood was perching very upright on the nest's rim, facing inward. It remained there while I regarded it a yard away, but as I turned to go it flew off, leaving the nest empty. Soon afterward, I found one of the fledglings in the shrubbery beside the pond, where it hopped and flitted with agility from branch to branch, well able to elude me. While perching, it wagged its tail up and down to balance itself. The accompanying parents scolded me with harsh, grating churrs and screams suggestive of agony. The older nestling had remained in the nest for eighteen days, the other two for seventeen days.

For the next ten days, I failed to see the young Donacobiuses when I visited the pond, although their parents complained loudly when I came in view. Finally, when the juveniles were twenty-six and twenty-seven days old, I found two of them resting in bushes above the pond's edge in bright sunshine while the adults brought them food. One of the young birds busily preened and scratched, just as the parents so often did. Aside from their smaller size, the most conspicuous difference between these juveniles and the adults was now the light stripe behind each eye of the former. They already had prominent white patches on their wings, and the ends of all but their central tail feathers were extensively white.

The long incubation period of seventeen days and the nestling period of seventeen or eighteen days of the Donacobius are noteworthy. From my own observations and published records (especially Bent 1948) it appears that in the species of the mockingbird family inhabiting the United States the incubation period is usually twelve to fourteen days, rarely shorter or longer. Nestling periods fall mostly within the range of eleven to fourteen days, although Palmer's Thrasher has been reported to remain in the nest for eighteen days. In the color of the inside of the nestling's mouth, the Donacobius also differs from other Mimidae for which information is available, for their mouths are yellow or orange-yellow without dark spots on the tongue. The Donacobius' lack of natal down also appears to be rather exceptional in the family, although the newly hatched Crissal Thrasher is said to bear only "the faintest suggestion of down on head and back." Apparently Donacobius, despite its thrasher- or mockingbirdlike form, is not closely related to the Mimidae. Recent anatomical studies suggest that it may be more closely related to the wrens; but no undoubted wren that I know builds a cup-shaped nest without a roof unless it is sheltered in a birdhouse, hole, or cranny of some sort. Donacobius is in many ways an exceptional bird whose acquaintance is well worth the pains that I took to cultivate it. With only a single nest, I did not learn everything of importance about its breeding. Along the marshy shore of an oxbow lake in Manu National Park in southeastern Peru in a later year, Kiltie and Fitzpatrick (1984) found eighteen territorial groups, each of two to four grown birds. The larger groups consisted of a mated pair with one or two recent young who helped the parents defend their territory, guard the nest, and feed the nestlings. The Black-capped Donacobius, as the bird is now called, is a cooperative breeder.

21. *The Message of Birds*

One of the encouraging developments of our time is the steadily growing interest in birds. Were this merely a resurgence of the old collector's mania for boxes full of eggs and for stuffed skins neatly laid in cabinet drawers or else mounted with staring glass eyes, the new preoccupation with birds would be deplorable rather than heartening. Happily, the recent growth of interest has taken a more commendable direction. It is, above all, appreciation of the living bird in its natural setting. It is bird watching, not bird collecting. It sends the enthusiast through fields and forests and marshlands, along the inland waterways and the seashores, equipped with binocular, notebook, and camera rather than with a gun, and it brings him or her back laden with notes, memories, and photographs rather than with limp feathered corpses that have lost the warm vitality which is the essence of a bird. Or those less strenuous and mobile in the pursuit of their hobby draw the birds to their dooryards and gardens by supplying food and planting shrubbery that offers shelter and sites for nests. This widespread interest in the living bird has stimulated the formation of numerous clubs and societies, with meetings and publications devoted to the discussion of observations and experiences with birds. The presses pour out ever more guides to facilitate recognition of birds in the field and countless volumes on the habits of birds and the adventures of those who go in search of them.

What is the significance of all this preoccupation with the feathered kind? What is sought by the thousands of bird watchers who spend long hours marching along rural lanes and trudging over the fields in all kinds of weather, returning home with no tangible trophies of their strenuous quest, yet feeling richly rewarded for their effort? What of the inmost nature of the seeker does all this seeking reveal? To what extent do they find what they seek?

It is easy to point out some of the attractions of bird watching and to show how it fills certain obvious gaps in the lives of people who dwell and work in artificial surroundings, immersed in the worrying complexities of modern life. There is the aesthetic appeal of creatures beautiful in form and color, swift and graceful in movement, gifted with melodious voices. There is the perennial excitement of hunting for the hidden and the exhilaration of stumbling upon the unexpected, for these mobile winged creatures are here today and far away tomorrow and there is always the possibility of meeting in one's own shade trees some rare bird unknown to one's neighbors. There is the advantage of combining with necessary exercise a pursuit that sharpens the senses of sight and hearing, exercises the mind, and stimulates the fancy. There is the charm of a quest that not only leads one into the fairest of rural and sylvan scenes but gives a point to all one's rambles, an added zest to every excursion. For those endowed with enough patience, there are the thrills of discovering cunningly hidden nests, the insights to be won by the self-effacing observation of the devoted parent birds. Systematically cultivated, bird watching brings wider knowledge of the natural world and deeper understanding of fundamental biological principles.

It is widely recognized that men and women, girls and boys, are led by these advantages to become watchers of birds. But that they are often, perhaps usually, drawn to this pursuit by a more profound and subtle impulse, so deeply embedded in the spirit that they themselves may be unaware of it, is frequently overlooked. Society, always jealous of the minds and loyalties of its members, tries insidiously to absorb them wholly into itself. This tendency, already evident in primitive tribes, takes subtler modes in modern civilization, where, on the one hand, people have fewer contacts with the natural world that sustains human no less than all other life and, on the other hand, the narrowing of vision for which science is at least indirectly responsible turns their thoughts and aspirations away from that vast unseen realm that surrounds and penetrates the small segment of reality revealed by our senses and measured by our instruments. But deep within us is something that refuses to be satisfied with the amenities of a mechanical civilization, intercourse with our fellows, and the expanding discoveries of positive science—something that society strives in vain to domesticate and place wholly at the service of its vast, ponderous, insatiable organism. We instinctively yearn for contact with something that envelops and transcends the purely human world which so loudly asseverates its own adequacy, so insistently presses its claim upon our bodies and our spirits.

The churches, once the chief stairways by which men sought to ascend from the humdrum human world to a transcendent realm that gave it higher significance, have lost much of their old appeal and authority. One reason for this decline in their liberating influence is the obsolescence of their symbols. For religion, which speaks to men of things never seen or heard or touched with the hands, must enter human minds by means of signs; yet the symbolism adequate for one generation becomes fantastic to another, whose ideas and practical experiences flow in different channels. The yearning toward a larger sphere, no longer satisfied by conventional religion, seeks fulfillment in other directions. Often it turns to nature, so much older, wider, more stable and dependable than our feverish human societies, so silent and enigmatic that one may regard it, if not as that ultimate reality toward which the religious aspiration impels us, at least as a more adequate and revealing symbol of that reality than any which the human mind has devised.

I do not intend to suggest that this

consideration is explicitly present in the minds of the great mass of enthusiasts who spend so much of their leisure time combing the countryside with binoculars to spy the earliest returning migrants or to add yet another rare species to their list of birds seen. Probably few of them have ever thought of their hobby in this light, and some would strenuously reject the implication that a recondite or spiritual motive underlies their pursuit. Nor do I intend to create the impression that I regard bird watching as a truer or more adequate outlet for the impulse to which I allude than botanizing, star gazing, or any other form of dedicated association with nature. I selected bird watching chiefly because of late it has attracted so many zealous votaries that it is coming to occupy an important place in modern culture. My contention is simply that many people devote their spare time to birds rather than to stamp collecting, or golf, or the theater, or any other occupation concerned solely with the activities or artifacts of man because, among other things, it is one mode of contact with that larger, embracing realm in which humanity is a late arrival.

Bird watching, then, is an indication of our human need to reach out beyond the narrow confines of human society and establish contact with a wider, more inclusive order of being. Birds, which in some measure satisfy this need, are like ourselves segments of this larger world and may be regarded as symbols or modes of expression of a basic reality of which the whole visible world is the manifestation. In seeking birds, we become aware of certain facets of our own nature too often overlooked. But can we go farther than this and by means of the birds that are the immediate objects of our quest gain some insight into the movement that created them? Every search reveals something of the character of the seeker and, likewise, in the measure that it is successful, it gives us glimpses of the nature of its object. The first point has received sufficient attention. Let us turn now to the second.

Whatever the character of the creative energy that has shaped our planet and covered it with life, whatever the goal toward which it may be directed, birds seem well fitted to reveal these mysteries to us, for on Earth's surface they are so abundant and widespread that we must regard them as no accidental or aberrant outcome of the creative process. On the contrary, they are so prominent a part of the living community that we must view them as a major expression, at an advanced level, of the energy that produced and sustains life. Compared with the vegetation that covers the planet's more benign regions with a green mantle, birds account for a small fraction of the mass of living matter, yet of the animal kingdom they are one of the most flourishing branches. Wherever he wanders over Earth's surface, birds claim the attention of the observant traveler more than any other group of animals, save possibly the much more numerous but individually smaller insects. By voice, color, and movement, these vivacious creatures of air and light reveal themselves to people more freely than the usually bigger but duller, more silent, and often nocturnal mammals and far more than reptiles, amphibia, or any humbler form of terrestrial life. Without the voices of birds, magnificent tropical forests would be silent as the desert and seem almost devoid of animate creatures. On the great grasslands and over arid wastes, the soaring forms of birds remind us of the omnipresence of life. During the short arctic summer, the subpolar tundras teem with nesting birds. Even on the high seas, hundreds of miles from shore, birds, rather than fish or cetaceans, are the animals most frequently glimpsed from the deck of a ship.

Under what aspects do birds present themselves to us? Most obviously, that of beauty, and this is what chiefly attracts us to them. Many are arrayed in the most brilliant colors, and even on those more soberly attired, dull shades are so soft and warm, blended so delicately in such intricate patterns, that when we contemplate them attentively we may ask whether these unpretentious sparrows, nightjars, and quails are not more beautiful than the gaudiest of the macaws and tanagers. But the aesthetic appeal of birds owes as much to delicacy of form, to the soft loveliness of feathers, and to swift yet graceful movement as to coloration. And they charm us through our sense of hearing no less than that of sight, for birds are nature's musicians. To all this we must add the beauty of nests so variously and skillfully wrought and the elegance of eggs in their shapely forms and endless diversity of shade and ornamentation. Birds account for no mean share of a privileged planet's beauty.

The second aspect under which birds present themselves to us is that of friendliness. In regions where winter's dearth or prolonged drought does not scatter the feathered tribes or send them afar in search of food or warmth, a substantial proportion of all the birds live in pairs throughout the year. Since for many months their reproductive impulses may slumber profoundly, it is evident that something other than sexual attraction binds the male and female together—personal affection or something similar. The sociability or friendliness of birds often leads them to join in flocks which are not incompatible with the maintenance of the bond between mates, as one becomes aware when he watches parrots flying overhead in a great company made up of couples flying side by side.

These modes of association are familiar enough among ourselves, but what appears strange to the human observer who pauses to reflect, what sets birds sharply apart from men, is the prevailing absence of hostility between species. In general, we featherless bipeds are feared and shunned by nearly every kind of terrestrial animal larger than insects, except the few that we have domesticated for our own purposes. But among birds, individuals of diverse species band together in a friendly company. One meets such mixed flocks in northern woodlands, but they are more important in the economy of the birds of tropical rain forests. Here, where the ecological situation does not favor the formation of large companies of a single species, birds with diverse methods of foraging hunt together. The companionship of such motley flocks is perhaps a psychic need of birds cut off from close association with others of their own kind. Since these mixed groups are composed not only of distinct species but of members of different families and orders, their equivalent among mammals would be a party comprising men, deer, antelopes, bears, rabbits, foxes, and so on—one or a few of each kind—keeping close company through much of the day with never a serious conflict between two individuals. The very notion of such a band strikes us as fantastic, the stuff of messianic visions or of fairy tales.

The third aspect of bird life that claims our attention is its orderliness. In some manner of which, despite various theories, we have inadequate understanding, birds find their way over vast stretches of Earth's surface to the very same garden or meadow where they nested or wintered the preceding year. Their arrivals and departures are so regular that one might suppose that they time their journeys by observing the positions of the celestial bodies. Then, too, order is brought into their lives by the widespread system of claiming territories or circumscribed plots of land, where each pair rears its family with a minimum of interference from others of their species who, on the whole, respect their neighbors' boundaries. In watching the rearing of a brood of young birds, we behold not only an admirable exhibition of parental devotion but marvelous coordination of the activities of the two parents and the reactions of their helpless offspring.

Such close interlocking of the behavior of the several members of the family group is an expression of innate or instinctive modes of behavior rather than of learned or rationally directed conduct, as in ourselves. We hesitate to affirm that the parent bird attends its nestlings or respects the boundaries of its neighbor's territory from a sense of duty or in obedience to the dictates of conscience. Yet this beneficent regularity of behavior, promoting the perpetuation of life and the prosperity of individuals, is the end toward which the greater part of our explicit human morality is directed, so that, wherever we encounter such ordered patterns of activity, we are constrained to recognize a moralness, or protomorality, of which our self-conscious morality is evidently a further development.

Birds, then, reveal to us beauty, friendliness, and an orderliness of behavior such as that to which human morality aspires without always achieving—not in one species only, or in one small part of the planet's surface, but among thousands of kinds which together comprise the most conspicuous division of animal life on Earth's continents and islands. We must regard them as a highly important and characteristic expression of whatever force shaped this planet and covered it with life, as a revelation of the direction in which the world process is moving. If we accept the birds as a true indication of this direction, we cannot deny that it is toward the realization of beauty, of friendship or love, and of a moral order—that is, toward the realization of the values that men have traditionally esteemed most highly, with the exception of truth or disinterested knowledge. We need not pause here to examine the nature of the creative energy, but perhaps it is more important, as it seems easier, for us to discover whither this power is tending than what it essentially is. The growing interest in birds that suggested this train of reflections appears to result from the striving of the spark of creative energy immanent in each of us individually to realize its affinity, or to establish harmony, with that great source whence it sprang.

Thus this reaching toward birds, as toward other aspects of the natural world, reveals something not only about our inmost selves but also about the larger whole to which we belong. The ancients believed that by trained observation of the flight and other activities of birds they could interpret the will of the gods and foretell future events. In somewhat the same spirit, we of a less credulous age may, by sympathy with birds, deepen our insight into the cosmic process of which they and we are products.

22. A Mockingbird in Blue

As I walked along a steep, bushy slope high in the mountains of western Guatemala in the late dusk of a November evening, I was arrested by an amazing medley of bird notes issuing from one of the low, scrubby, second-growth oaks scattered over the hillside. From amid the dark leaves came a rapid flow of monosyllables, now a shrill squeak, now a whistle, now a guttural croak, all intermixed in the most surprising fashion. I maneuvered around, trying to glimpse the author of these startling utterances; but the light was already too dim to distinguish anything among the dense foliage. I hurried back to supper with a mystery revolving in my mind.

The mystery was solved sooner than I expected. The following morning, I climbed to the summit of the ridge that rose steeply behind the house where I was a guest. There, in the undergrowth beneath pine and oak trees, I glimpsed a pair of finches with olive backs, black heads with a narrow white stripe over the center of the crown, and bright yellow throats. I saw them only fleetingly before they vanished into the undergrowth, and I tried hard to follow and learn more about them. While engaged in a fruitless search for these shy birds—my first Yellow-throated Brush-Finches—a repetition of the performance that had so amazed me on the preceding evening came from an unseen source. I forgot the yellow-throated birds in my eagerness to clear up a mystery one night old.

The abrupt alternation of high and low monosyllables was the outstanding feature of the vocal hodgepodge that now claimed my attention. I distinguished a little peep, a short, clear whistle, a churring note as of a woodpecker, a guttural *chuck,* a brief screech, attempts at warbles and trills. The songster, if so he might by courtesy be called, was apparently trying to imitate notes that he had heard from other birds, but I had not yet been in the Guatemalan highlands long enough to recognize the species that he was attempting to mimic. The effect was amusing, fantastic, pleasant in its way, certainly not musical or harmonious. Following the sound, I finally glimpsed a big blue-and-white bird who promptly vanished. Pursuing with utmost caution, I found him behind a low ridge of earth where I could see the leaves that he tossed into the air while he himself was screened from view.

I dared not approach nearer. It was exasperating to stand and watch brown leaves rise and fall above the ridge, fearing that the hidden source of these movements would fly away before I had seen enough to identify him. However, after rummaging in the ground litter for a while, the bird obligingly rose to a low perch and continued to chatter while I enjoyed an excellent view of him. Nearly

Blue-and-White Mockingbird

11 inches (28 centimeters) long, he had slaty blue upper plumage, a black mask over his face and ear coverts, and pure white under plumage dulling to grayish blue on his sides. His slender bill was black, his eyes dark, his legs and toes blackish. From the description I then wrote, I later identified this beautiful bird as the Blue-and-White Mockingbird, a species found from the Mexican state of Chiapas through Guatemala to Honduras and El Salvador, in dense thickets and open woods with abundant undergrowth, from about 2,000 to 9,000 feet (610 to 2,745 meters) above sea level.

When I returned to the Sierra de Tecpán in February of a later year, these birds had already paired. Except when he sang, I could not distinguish the male blue mockingbird from his mate. One pair roosted in a dense tangle of shrubs and blackberry canes beneath tall Alder trees. In the waning light of evening, the male mingled his quaint medley of sweet and throaty notes with the calls of the Whip-poor-will until he fell asleep and left the latter to call alone; in the morning, he resumed his half-melodious chatter as soon as he awoke. The green berries of a nearby viburnum bush provided the first course of the pair's breakfast. After swallowing whole a number of these little fruits—which are distressingly astringent to the human mouth—they dropped to the ground to find the remainder of their meal beneath dead leaves, tossing them aside with their bills, rarely scratching with their feet. This was how they found most of the small invertebrates that were their chief food. Little patches bare of leaves and litter marked the spots where the mockingbirds had foraged.

Toward the end of March, I learned that I had been unjust to the blue mockingbird by thinking of the medley which first made me aware of his kind as his song. When he delivered that curious mixture of sounds, he was merely amusing himself or chattering to his partner, not trying to display his musical genius. Now he began to show the whole mountain what he could do as a songster. He repeated over and over a loud, ringing, liquid trill that was most beautifully modulated. One of his neighbors delivered a series of low, soft whistles so ventriloquial that it took me long to locate the bird among low bushes. Both of these so different strains seemed to be original rather than imitative. I surmised that in its nesting season this bird would develop into a peerless songster.

My faith in the blue mockingbird's vocal ability was amply confirmed two months later. A male who entertained me while from a blind I watched his mate incubate toward the end of May was an incomparable songster who could produce with equal ease notes ranging from a deep, mellow whistle to a light, airy trill. He had, like the Yellow-tailed Oriole, a great variety of short musical phrases, each of which he repeated until he tired of it, then chose another. When at his best, he was not really a mockingbird— that role among the birds of the Sierra was ably filled by the Black Thrush—and he rarely imitated other birds. During a whole day I heard him borrow only one verse, that of the Whip-poor-will, and the likeness was so exact that I was certain that he copied this phrase from the nocturnal bird and did not by coincidence invent it himself.

Sometimes this mockingbird sang while concealed in the impenetrable thicket, sometimes he rose to the topmost twig of a tall pine tree to pour forth his exquisite notes to all the mountainside. He began the day with his choicest selections and during the first hour performed in his most inspired manner. Thereafter, he seemed to become bored with classic music and turned to nonsense songs to pass the time, introducing many harsh and churring notes into his medley for variety. In the early morning his recital had been more pleasing than that of the gray mockingbirds because of the absence of all harsh interludes; now he descended to the level of his gray relations and through the remainder of the day much unmelodious chatter was interspersed among his charming verses. The dark, rainy weather of this season failed to depress his spirits, as it sometimes did mine. On the contrary, he seemed to enjoy the mist and rain.

The only notes I heard from a Blue-and-White Mockingbird definitely known to be a female were low and squeaky or else an oft-repeated guttural *chuck*.

On the Sierra de Tecpán, between 8,000 and 9,000 feet (2,440 and 2,745 meters) above sea level, the blue mockingbirds must have started their earliest nests late in April, for two full-grown birds that I saw following their parents in early June could not have been hatched from eggs laid later than the first of May. However, I did not find nests until after the rains began in mid May, when the builders of these nests were just starting to incubate. Between this date and early July, I discovered five nests. Like the Ruddy-capped Nightingale-Thrush and the Chestnut-capped Brush-Finch, which also foraged on the ground, the mockingbirds nested chiefly in the first two months of the rainy season. English ornithologist Osbert Salvin (Salvin and Godman 1879), who had studied the habits of this mockingbird many years earlier, found them nesting from late May until September. His observations were made several thousand feet lower than the Sierra de Tecpán, and at lower altitudes in Guatemala many birds breed farther into the wet season than on these cool heights. The last nest that I found on the Sierra was deserted toward the middle of July while it still held two eggs, apparently from no other cause than the lateness of the season.

Nests of the Blue-and-White Mockingbirds were situated in dense thickets, more rarely in a sapling beneath open woods, from 4 to 15 feet (1.2 to 4.5 meters) above the ground. The female built alone. First, she made a loose and bulky framework of long, coarse sticks which she spread apart in the center to receive a shallow, thick-walled cup of closely matted fibrous roots, neatly arranged. Probably she worked hard to pull up the roots, which she brought doubled up in her bill and tucked into place while she stood on the sticks beside the cup. Then she sat in the nest, turning to face in different directions while she shaped it with feet and breast. Her beautiful, immaculate, bright blue eggs were conspicuous in the dark and shallow bowl. Each of my five nests contained only two eggs or nestlings, although Salvin found three in most of his nests.

The nest that I watched was low in a dense tangle of bushes and blackberry canes, beside a rivulet that flowed with a loud babble through a deep and narrow valley between steep slopes covered with bushy growth and scattered trees. In an attempt to mark the female with a spot of paint to distinguish her with certainty from her mate, I stuck a small paintbrush with its end soaked in vermilion enamel into the rootlets and sticks of the nest, projecting over the eggs. Then I hurried into the little tent that I had already set near the nest to watch. After hesitating long and voicing many throaty *chuck*'s, the female finally returned to her eggs. Finding the strange object in her way, she promptly seized the end of the improvised paintbrush, pulled it from between the rootlets, and flew away with it. When she had not returned in more than an hour, I removed the blind and departed.

The next day, finding that the mockingbird had not abandoned her eggs, I tried again to mark her, this time fastening the brush to the nest so securely that she could not easily remove it. The result was that the eggs remained cold for more than two hours, although this time I went out of the valley, leaving no blind to arouse the mockingbird's suspicions. It appeared that they would desert the nest rather than stain their beautiful white breasts. I have often succeeded in marking flycatchers, ovenbirds, cuckoos, and other birds by this method to help me distinguish the members of a pair alike in plumage, but I have found songbirds of all kinds so careful not to stain their feathers that I long ago stopped trying to make them do so.

As it turned out, my inability to distinguish the female by her plumage was not detrimental to my study, for I could always recognize her partner by his voice. On the afternoon of the day of my second failure to mark her, I began to watch the nest from the blind and continued until nightfall; then I resumed my vigil before the following dawn and continued until the hour at which I had started on the preceding day. It was then late in May; throughout the afternoon, rain fell continuously, sometimes in a hard downpour, sometimes in a light shower, and during my morning watch frequent showers alternated with brief intervals of sunshine. Nevertheless, the male sang during every one of his partner's sessions on the nest; since she was restless, this means that throughout the day not a single half-hour passed without his cheerful notes.

As the dim, rainy dawn broke over the sodden mountain, the Black Thrushes awoke to sing and after a few minutes were joined by the Ruddy-capped Nightingale-Thrushes. The pensive, ethereal notes of the nightingale-thrushes, heard against the background formed by the full, powerful voices of the Black Thrushes, made a wonderful dawn chorus—finer than such a drab day deserved. It seemed an incongruity when the Chestnut-capped Brush-Finch, whose mate was nesting across the rivulet, intruded his squeaky voice among those of the better songsters. The thrushes' chorus had begun to wane before the mockingbird joined it with his clear, mellow notes. Although he passed much of the day within hearing of the nest, sometimes he went off on long excursions by himself until his notes faded away in the distance. The female usually remained on her eggs until he returned and his voice sounded nearby, then left them to fly beyond sight with him. He almost always accompanied her when she returned to her nest, but he never approached nearer than 6 feet (2 meters) from it.

The mockingbird who covered the eggs never revealed any ability to sing, not even so much as the female Gray Catbird, her distant cousin. While sitting in the nest, she did not once sing softly in response to her mate, as do the female Yellow-tailed Oriole and the female Melodious Blackbird, nor did I hear her join the male in song when she went off to forage in his company. While incubating she was very silent; I heard her only once when she answered her mate's song with a few low, squeaky notes.

For so large a bird, the female Blue-and-White Mockingbird's periods on the nest were short. During the fourteen hours that I watched, she took twenty-eight sessions on the eggs ranging from 8 to 42 minutes, with an average of 20.8. An equal number of recesses ranged from 1 to 23 minutes and averaged 7.1 minutes. She devoted 74.6 percent of the day to incubation. It is of interest to compare her behavior in the forenoon, when only brief, light showers fell although the sky was cloudy and the air held a penetrating chill, and in the afternoon, when rain fell strongly and steadily. The rain did not cause the mockingbird to prolong her periods on the nest, which in the afternoon averaged less than a minute longer than in the forenoon. But the afternoon rain shortened her recesses, which were never more than 7 minutes and sometimes only 1, and on the average only one-third as long as in the morning.

This was quite different from the behavior of a Slate-throated Redstart whom I had watched on a similar afternoon a week earlier; while rain fell, the redstart substantially lengthened her absences. The different behavior of these two birds in the same kind of weather appeared to be related to their diverse ways of foraging. By washing insects from the air, the rain made it more difficult for the redstart, who catches much of her food on the wing, to satisfy her needs, but it did not adversely affect the mockingbird's search for small creatures on the ground or for berries on shrubs. On the contrary, since small invertebrates tend to come to the surface in wet weather, the rain probably made it easier for the mockingbird to find them. Apparently, this is why Blue-and-White Mockingbirds and other ground foragers nested chiefly in the rainy season, whereas the redstarts and most other birds on the Sierra de Tecpán began no new nests after the rains returned.

When I ended my long vigil early in the afternoon of May 26, I went up to the mockingbirds' nest and found one egg already pipped. Placing the other at my ear, I heard tapping in it. By the following day, both eggs had hatched. Unfortunately, they had been laid before I found the nest, so I could not learn the length of the incubation period. The sightless, dusky-skinned nestlings were sparsely covered with long, soft, blackish down. The insides of their mouths were bright orange-yellow.

Three minutes after I resumed my watch at this nest, the female returned to brood the two-day-old nestlings. After another three minutes, their father arrived with a billful of food. When their mother opened her mouth widely, he stuck it well down into her throat. Of the several pieces that he brought, she dropped one, which he recovered, then hesitated to place it again in the mouth that she opened to receive it. He looked for the mouth of a nestling to whom he might deliver the food directly, but their mother had them too well covered, and with apparent reluctance he finally relinquished the last item to her. She stepped backward to the nest's rim, fed both nestlings, then resumed brooding. Soon she left for a brief recess, returning from which she brought food, gave it to the nestlings, and brooded again. As she was leaving once more, her mate, coming with food, met her among the bushes about 3 yards (2.7 meters) from the nest and passed to her what he held in his bill. Again she dropped a piece and, following it to the ground, found it after searching for about two minutes. Then she returned to the nest, fed the young, and brooded them.

In two and a quarter hours, each parent brought food four times. The female brooded six times, for intervals of from 11 to 21 minutes, with intervening absences of from 2 to 18 minutes. During the nestlings' first days, their father never fed them in the absence of their mother; if he arrived with food while she was away, he hopped among the bushes a few yards from the nest until she returned and received it from him. Later, however, he fed the nestlings while his mate was absent and removed their droppings. As he flitted toward the nest with a full bill, he sometimes sang in an undertone. Now that family duties weighed upon him, he had little time for the elaborate vocal performances of the days when his partner incubated.

After watching the mockingbirds attend their nestlings for a while, I decided to give these parents some problems to solve. While they were beyond view, I completely covered their nest with a large, downy leaf of a shrubby *Senecio*. Soon returning with food in her bill, the female hopped all around the nest, trying to look under the leaf while she constantly repeated a low, throaty note. After three minutes of this inspection, she cleared her bill for action by swallowing what it contained, picked up the leaf, and easily carried it from the nest, then promptly returned to brood. While she covered the nestlings, her mate brought food and passed it to her for delivery to them, as before. When she left, I emerged from the blind and spread a white handkerchief over the nest. This time the parents returned together, both with full bills. The female, who was slightly in advance of her partner, promptly swallowed what she carried and tugged at the handkerchief, which caught on the sticks in the nest's foundation, was difficult to remove, and seemed to frighten her. She retreated a short way and the male came forward, swallowed the food in his mouth, and pulled the handkerchief until it was clear of the nest, then dragged it about a foot away. Here he hopped around it, at times spreading his wings, jerking it and trying to move it farther from the nest, but it had caught on the thorns of a blackberry bush and he eventually abandoned his efforts to release it.

After the mother had warmed her nestlings again, I transferred them from their proper nursery to my cap, which I had made into a substitute nest among the bushes nearby. Soon the female returned, approaching from the side of the nest opposite the cap, this time without food. She peered into the empty nest and poked at the bottom with her bill. Although clearly perplexed, she sat in the nest as though to brood. Here she did not feel right, constantly shifted about, and rose to look beneath herself. A loose stick annoyed her. She tried to push it down into the nest's rim where it belonged but could not arrange it to her satisfaction, so after sitting for less than two minutes she carried it away.

During the female's absence, the male several times approached the nest singing, with food in his bill; but because at this period he would not go quite to the nest unless his mate was present he did not learn that the nestlings were missing. When, after half an hour, the mother did not return, I began to fear that the two-day-old nestlings would suffer from exposure, for the sky was overcast and the air damp and chilly. Again leaving the blind, I found the chicks quite cold to the touch, hardly able to move. I warmed them inside my shirt until they recovered their vitality, then replaced them in their nest, took my cap, and hurried away. The mother returned to brood them, and they suffered no ill consequences from their exposure.

The parent mockingbirds, so quick to take effective action when their nest was concealed beneath a leaf or a cloth, failed to meet the situation created by the removal of their offspring to a makeshift nest in plain view only a few feet from the real one. Indeed, they hardly even searched for them. If I had kept the nestlings in the cap much longer, they might have died of exposure. In extenuation of the parents' negligence, we should remember that when nestlings vanish from a nest before they are feathered they are nearly always irretrievably lost.

A week later, I repeated my last experiment, this time placing the cap with the young mockingbirds a yard from the nest on the side opposite that by which the parents always approached and departed. Their mother was the first to return with food, and on finding the nest empty she flitted through the surrounding bushes, repeating her usual throaty *chuck* and going thrice to look into the empty bowl. On leaving the nest after the fourth inspection, she found her missing offspring, only one minute after her reappearance; but instead of feeding them she went again to the nest, then back to the cap, then once more to the nest and back again to the cap. After four minutes, she swallowed the food in her bill, then continued to circulate through the bushes a few minutes more, repeating her throaty note every few seconds before she settled down to brood the empty nest. Here she sat for six minutes, then hopped out and vigorously preened her plumage in a neighboring bush.

While she was engaged in this occupation, her partner approached singing, with food for the nestlings in his bill, and passed her on the way to the nest. Finding it empty, he lowered his head above it in a most comical fashion, as a nearsighted person might do in similar circumstances. Convinced that the nestlings were no longer present, he flitted through the bushes and within two minutes of his return discovered them in the cap. Several times he advanced toward the strange object and retreated from it, but soon he overcame his distrust and fed a nestling, then went away, singing. As in the episode of the handkerchief, he proved himself more capable than his mate.

When the male brought food once more, from force of habit he went first to the empty nest, then immediately to the cap. The female soon followed his lead, and so long as I left the nestlings in the cap both parents went directly to feed them. But their mother, whose duty it was to warm them, would not brood them, although several times she appeared to be on the point of doing so. The young mockingbirds, now nine days old, had developed a surprising capacity to regulate their body temperature. Although considerable areas of their skins were not yet covered by their sprouting plumage and the afternoon was so cool that my nose, ears, and fingers soon felt as cold as the steel blade of my machete, they remained quite warm during their two hours' exposure. Now, when they heard a parent approaching, they rapidly repeated a soft peep which helped guide the adults to their new location.

Parent Blue-and-White Mockingbirds, excessively shy and retiring, gave little evidence of anxiety when their nestlings appeared to be in danger. If their mother happened to be brooding them when I approached, she always stole away before I could come close enough to glimpse her on the nest. While I remained at the nest, both parents lurked some distance away in the thicket, where they hopped around in silence or voiced throaty notes, never emitting any cries of distress or attempting either to attack or to lure me away. Another female mockingbird was braver or perhaps felt more secure in her higher nest, where she would remain sitting quietly while I passed beneath her. She even stuck to her post, head raised and alert, while I set up a ladder and climbed to the second step, but when I rose higher she fled and preserved a most respectful distance while I examined the ten-day-old nestling that she had been brooding.

The aloofness of the Blue-and-White Mockingbirds contrasted strongly with the behavior of Brown Thrashers whose nests are disturbed by humans and with that of a pair of Gray Catbirds whose nest I watched in Maryland. Unlike the blue mockingbirds, the latter failed to remove a handkerchief or a leaf that covered their nest; but they did not lack courage, and when I touched their nestlings the female alighted on the back of my hand to peck it while the male buffeted my head from behind, and both repeated mewing notes of distress. The catbirds nested in a suburban hedge, where they saw many people and learned that they had little to fear from them; the mockingbirds dwelt in a secluded valley in a region where men were mostly hostile to birds.

When twelve days old, the mockingbirds were well clothed in plumage. Now, instead of stretching up their open mouths for food when I bent over them, as they had done until the preceding day, they crouched down as I approached. When two weeks old, one cried out and tried to bite my hand. The other was more pacific. When I lifted them from the nest for inspection, the former tried to escape. Despite vigorous flapping of wings, it could fly only downward and soon landed on the ground, over which it hopped with alacrity. Returned to the nest, the young birds stayed quietly at home. The parents retained their aloofness to the very end; even the cries of frightened nestlings could not draw them to confront me.

As long as the young remained in the nest, their mother brooded them by night and while rain fell. Those to whom I devoted most attention left when fourteen and fifteen days old. Perhaps they would have lingered a day or two more if I had not removed them for examination.

When they departed the nest, the young mockingbirds were everywhere, except on the abdomen, so dark a gray as to be almost black. But the gray feathers of throat and breast were sprinkled with white, signs of the approaching whiteness of these regions. The white feathers of the abdomen had gray tips. The bill was light gray with white edges, the inside of the mouth orange-yellow, the eyes dark brown. This juvenile attire was worn only a short while. A young mockingbird that I noticed in early August, my attention drawn to it by the characteristic throaty *chuck*, had chin, throat, and breast already largely white with a few conspicuous traces of gray. Every mockingbird that I saw after early September was in full adult plumage. The families dispersed after the young birds could care for themselves. During the last months of the year, I saw solitary individuals so frequently that I doubt whether these mockingbirds, like so many other birds permanently resident in the tropics, remain constantly in pairs. After the nesting season, males ceased to sing melodiously but continued to voice the strange medleys that first drew my attention to them.

23. *The Lonely Vanguard*

From ancient times, the beauty of the nocturnal sky, the precision and constancy of the movements of the Sun and stars, have inspired thoughtful men with wonder and admiration. Beings that endured unchanged for long ages, that moved across the firmament without collision or strife with neighbors, were revered as immortal gods by ancient sages. Only the planets, appearing to move now forward and now backward among the constellations, troubled certain Greek philosophers by their ungentlemanly inconstancy. Modern thinkers, aware that the stars are but huge masses of incandescent matter, that the planets and their satellites shine with borrowed light, may still believe, with Sir James Jeans, that the Universe was set in motion with mathematical precision by a master Intellect, that the order of the cosmos implies a divine Orderer, a God who is above all a cosmic planner or astrophysical engineer. Our very term "cosmos," from the Greek word for order or harmony, commemorates the ancient belief that the Universe is a beautifully ordered system.

When we turn from the greatest bodies to the least, we discover similar orderliness and durability. The smallest particles of matter, protons, neutrons, and electrons, appear to be even more enduring, more "immortal," than the stars which, in their inconceivable myriads, they compose; there is now evidence that the latter are born, grow old, and decay while their constituent matter surges through the Universe forever unless, perhaps, it is imprisoned in one of the astronomers' black holes. The chemist can predict the interaction of two substances with hardly less precision than the astronomer foretells the positions of the heavenly bodies. Although the same atom may in turn enter a variety of molecules and one chemical often disintegrates another, at the atomic level no destruction occurs in ordinary chemical reactions—the atoms simply change partners, like dancers. In the inorganic world, the so-called cosmic strife may be no more than a merry dance in which the inconceivably minute dancers restlessly alter their formations with never a pang or a regret and, perhaps, with gleams of pleasure for all concerned. Or, if these diminutive beings line up in crystals, they may repose unaltered for long ages in apparent bliss. It is as easy to believe in a God who is the Supreme Chemist as in a God who is the Master Astrophysicist.

About midway in size between the world of protons, electrons, and atoms and that of stars and galaxies stands a third realm, that of living beings, where a very different picture greets us. Regarded in isolation, a healthy organism is a supreme achievement of biological engineering. In it, compounds of great complexity are present in greater variety, interacting in more subtle ways and preserving a more self-sustained stability than one finds in any inorganic system of comparable size. If the existence of anything in the known Universe seems to point to an omniscient and omnipotent Creator, it is a living plant or animal.

But when, instead of focusing attention upon single organisms, we contemplate the whole realm of life, doubts assail us. The living world is, preeminently, the realm of strife. Daily myriad living things are torn piecemeal or devoured whole by other living things. Not only do the more advanced organisms exploit and consume those lower on the scale, which the former may suppose to have been made to serve them, but very often the lesser creatures afflict and destroy the greater, frequently in the form of multicellular parasites or microscopic pathogens. The flow of materials and energy through the living world, the so-called food chains, cannot, like the permutations of elements in chemical reactions, be regarded as a merry dance, devoid of pain and fear and possibly with pleasurable sensations for all the invisible participants. As we know only too well, the strife in the living world is accompanied by vast amounts of fear and pain, of hatred, anger, fury, and all the other violent passions that oppress us. Here is no neat exchange of partners, as when two compounds exchange atoms, but beautiful creatures are suddenly reduced to bloody messes from which we avert our offended vision. The living world is the realm of paradox, in which creatures that above all strive to preserve their integrity are inexorably doomed to disintegrate, if not by violence then by slow decay.

Although it is not too difficult to imagine a Divine Astrophysicist who rounded the celestial bodies and set them on their courses, or a Divine Chemist who formed the smallest particles and regulated their interactions, it is hard to conceive a Divine Biologist who has guided the evolution of life. Or if such a Being exist, he must be all power and cleverness, unleavened by compassion, who with infinite skill has formed a vast diversity of creatures that he does not lovingly protect from pain and horror. Such a God might be worshipped by those who admire unlimited power, but he would be repudiated by those who demand benevolence and love in the Deity they adore.

Perhaps it is a question of importance. All the matter in the Universe appears to be involved in physical and chemical processes, and a large proportion of it is aggregated into stars, planets, and other celestial objects. But, on the most liberal estimate, only an infinitesimal fraction of the stuff of the Universe is contained in living organisms; even on our planet, which seems to be more richly endowed with life than most, it occupies only a thin skin surrounding a huge, lifeless mass. Such a negligible quantity of matter may well be neglected by the ruling Power.

Life apparently arose from the strong tendency of the elements to combine in ever more complex aggregations wherever, as on the surface of a planet provided with water and an atmosphere and energized by radiation from a central Sun, they found themselves in an environment favorable for such developments. Eventually, some of the large molecules so resulting acquired the power to duplicate themselves by controlling the behavior of simpler substances that came within their sphere of influence, much as, on a more primitive level, a crystalline particle governs the alignment of atoms or molecules like that of which it is composed. With the acquisition by complex molecules of the power of reproducing themselves, life embarked upon its long, adventurous course.

The evolution of life depended, in the first place, upon chance alterations in the structure of the central or governing molecules of each aggregate of living substance—molecules which, in more advanced organisms, are situated in the chromosomes of the nuclei. Such alterations or mutations are caused by the incidence of radiation of high frequency, although chemical reactions are responsible for some of them. Evolution received a great impetus with the development of sexual reproduction. Itself no doubt a result of random changes in the living substance, sex greatly increased the element of chance in biological development. The union of two gametes, or sexual cells, is like shuffling together two more or less different decks of playing cards which are then dealt to the future progeny. The resulting "hand" which each receives may differ from that borne by any of its ancestors; it may give the progeny a substantial advantage in the game of life or, on the contrary, it may so handicap the creature that it succumbs. Sexual reproduction is essentially a form of gambling. Life may be said to have gambled its way upward from the primitive poverty of the most rudimentary organisms to the opulence of the higher plants and animals. But, as in all forms of gambling, the winners grow rich at the expense of the unfortunate losers. Albert Einstein, a mathematical physicist, could not believe that "God plays dice with the world," but chance enters into genic mutation and recombination hardly less than into a throw of dice. Life has had to advance by this crude and hazardous method because, without guidance, no better way was available to it.

The complement of mutation is selection, without which variation might have populated Earth with a multitude of inefficient, barely viable organisms. The idea of natural selection and its decisive role in the evolution of life grew out of observation of the results of artificial selection as practiced by breeders of plants and animals. But natural selection and selection by man operate by exactly opposite methods. The breeder of animals or plants works with an ideal of beauty or utility; selecting those individuals that show some advance in the direction of his ideal, he gives them special care and fosters their multiplication. Natural selection operates with no ideal or goal; it consists not so much in giving special advantages to superior individuals as in penalizing the inferior—inferior, above all, in ability to procure nourishment, escape enemies, resist environmental stress, or reproduce. Artificial selection is positive selection—choosing the best for protection and propagation. Natural selection is negative selection—picking out the weakest or least efficient for destruction. As has often been observed, the fitness implied by the phrase "survival of the fittest" means nothing more than fitness to survive and reproduce; it indicates no other excellent qualities. Since innumerable kinds of pestilent organisms and noxious parasites as well as fierce predators have proved their ability to survive at the expense and to the detriment of more amiable creatures, evolution by random mutations, sexual shuffling of genes, and natural selection has filled the world with organisms that perpetually harass and oppress other living things while contributing little to the beauty and joy of life.

In the tropics, the immense profusion of creatures that bite, sting, or otherwise irritate, that ruin one's possessions or destroy one's dwelling, repels many visitors from higher latitudes, while others are so captivated by the splendor and interest of tropical nature that they willingly endure all this and more. The wonder is that by such rude methods evolution has produced so much that is fine and lovable while it continues to fill with life every part of the planet, however harsh and forbidding, that is capable of supporting vital processes on however reduced a scale. Only a very strong impulse to raise existence to higher levels by whatever means are available can account for this.

When one surveys the living world broadly, observing what treasures of beauty, what marvels of perceptiveness, what heights of love and devotion the stuff of the Universe is capable of generating when molded into organic forms, one reflects with a pang what life might have become if its formation had been guided by some great and beneficent Power instead of being left to the rough, groping methods of organic evolution, gambling with destruction of the losers. Such a Creator might have covered our planet with loveliness and joy and sympathetic understanding, with beings of a magnificence, physical and spiritual, that we can hardly imagine, with none of the mean and loathsome and pestilent creatures that swarm upon it today. Above all, such a benevolent and skillful Creator would have avoided the horror of predation by making it unnecessary for any animal to strike down another to satisfy its hunger, and he would likewise have so regulated the reproduction of organisms that no species would have exceeded the ability of Earth to support it, with resulting deprivation and misery.

These two major evils, predation and overpopulation, are more closely associated than is generally recognized. Probably some animalcules began to devour others after they had become so numerous that primary sources of nourishment, such as nutrients dissolved in the primeval seas or algae capable of photosynthesis, no longer sufficed to support them. Species subject to heavy predation must reproduce abundantly to avoid extinction, as must those afflicted with fatal diseases or other causes of high mortality. If for any reason the predators diminish in numbers or efficiency, the rate of reproduction that they have forced upon their prey proves to be too high, with the result that the species relieved of this drain multiplies until it begins to exhaust its food supply and the surplus population succumbs to starvation. This situation has led ecologists to view predation as a blessing in disguise because it prevents the miseries that follow upon overpopulation. However, they overlook the probability that in the absence of predation the species preyed upon would not have developed such a high rate of reproduction. Animals show considerable ability to adjust their reproductive rate to their mortality, but such adjustment is achieved so slowly that any sudden increase or decrease in the death rate is likely to have disastrous consequences, the former perhaps leading to extinction, the latter to overcrowding with its attendant evils—to both of which man, no less than other animals, has been subject.

Man's religions are proof of the manifold evils that afflict a living world that has groped its way forward without guidance or regulation by some beneficent Power. We do not have these religions because gods or a God exists. It is precisely because there is no God, or at least none who cares lovingly for living creatures, that we have religions such as most of those that we know. For what are the majority of these religions but pleas for supernatural succor wrenched from the perplexed soul of a humanity tortured by pain and fear, from a humanity living precariously under the shadow of inevitable death? Whether we examine some primitive cult that strives to placate an angry or envious god by bloody sacrifices; or Buddhism, with its program for extinguishing an existence that it declares to be fundamentally painful; or Christianity, whose symbol, the cross, at the summit of every church proclaims to all the world that salvation is won through suffering, the conclusion is the same: we have these religions because no Deity at once powerful and benevolent cares about the living world.

I would not infer from this that a race ignorant of suffering, fear, and death would be devoid of religion. Every spiritual intelligence tries to relate itself ideally and in practice to the Whole of which it conceives itself to be a part, and such relatedness, devoutly cultivated, is the essence of religion. But in the happier world, religion would be a paean of joy rather than a cry of distress. Instead of fostering hope that death will be followed by some more blessed existence for which it can offer no convincing proof, it would promote the appreciation and loving care of a world that offers much to delight us. It would be a religion of thanksgiving rather than of salvation.

Although life has evolved without guidance by anything more ancient than itself, it is far from being unrelated to earlier phases of cosmic development. On the contrary, it represents a later or higher stage of a process that has pervaded the Universe at least from the beginning of the present cosmic era. Throughout visible space, we detect a movement of organization or harmonization which, acting on a grand scale, has rounded off the stars and set the planets revolving around them with majestic regularity. On a small scale, it has joined the ultimate particles—electrons, protons, and neutrons—into atoms which are sometimes pictured as miniature solar systems; it has joined these atoms into molecules of varying degrees of complexity; and it has lined up atoms and molecules in crystals that are often of great beauty and capable of enduring unchanged for long ages.

Carried to a higher level of complication and integration, the organizing tendency that forms atoms, molecules, and crystals brings forth living substance. It is hardly an exaggeration to say that life is the goal of matter, for wherever conditions are favorable to the delicately balanced vital processes, as on the surface of a planet neither too hot nor too cold, with abundant water and nutrient salts and an atmosphere that stabilizes temperature and supplies oxygen and carbon, matter enters the living state with a rush. The resulting superabundance of life is one of its chief troubles. If there were fewer living things to bump against, compete with, prey upon, or parasitize, life would be far more pleasant for all creatures. Like a developing child, life in its onward march has needed not only guidance and encouragement but likewise restraint and the moderation of its often excessive exuberance. Both the guidance and the restraint have been conspicuously lacking.

If the older constituents of the Universe from which life arose have failed to provide helpful guidance, they have at least supplied the impetus that carries it forward. If they have not beckoned it from above, they have pushed it upward from below, and this is the source of life's creative energy. They continue daily to give it substance and energy and a place to stand. Without the inorganic Universe, life could not exist in any form that we know. With this impetus from below, this lateral support, life continues, lonely and unguided, to grope its way forward into the unknown.

As is natural, man, who in certain ways marches at the van, seems more alone than any other creature. Despite his teeming billions and his dominant position on his planet, in his more thoughtful moods he sometimes feels devastatingly solitary. The gods to whom his ancestors looked for guidance and succor in times of distress, who when angry could be placated by gifts and expressions of submission and adulation, turned out to be but figments of their fertile imagination. With telescopes, huge radio antennae, and space probes he searches the Universe for indications of intelligent life beyond his own planet, thus far always in vain. Receiving no answers to his anguished pleas for assurance that some higher Power is cognizant of him, that he is not the only advanced intelligence to be found anywhere, he may conclude with Bertrand Russell that he lives precariously in a hostile Universe. As though he could exist in a hostile Universe! As though he were less a product of a universal creative process than the plants and animals around him and the stars above him!

Man's loneliness is largely the fault of himself and, ultimately, of the harsh conditions in which he, like other forms of life, evolved. Although primitive tribes often cultivated brotherly feelings toward nonhuman animals and even plants, more recent man, especially in the West, has fiercely resented the assertion that he belongs to the natural world—an attitude that has happily been changing in the present century. If a quite different animal of approximately equal intelligence and power, an animal with whom he could communicate freely and gain fresh knowledge and insights, shared Earth with man, he would probably try to exterminate it. If *Homo sapiens* had been less intolerant and belligerent, Neanderthal man and other less closely related tool-using primates might well be with us today.

Coexistence with other articulately intelligent animals, no matter how harmonious, or enlightening communication with extraterrestrial beings would not necessarily alleviate our loneliness, for they might be subject to all the doubts and hazards that oppress us. What man has long sought is some higher Being to whom he could look for help, guidance, and the assurance that he is not an accident of evolution but of transcendent importance. But, if we are at the forefront of the advance, how can we expect to find a guide or leader ahead of us? The time has arrived to outgrow the childlike need of a father, in heaven or elsewhere, and, at last maturing, make the best of our own great sources of strength. We must cultivate acute awareness of the power within us and its continuity with forces and processes pervading the Universe. Just as the topmost twig of a lofty tree, exposed to Sun and wind and rain with no shelter from surrounding foliage, might, if it could reflect upon its situation, feel lonely and abandoned unless it remembered that through roots and massive trunk the whole tree supported it, so, until we recognize our intimate connection with universal forces, we shall continue to feel abandoned and bewildered.

One continuous organizing movement sweeps in successive phases through the Universe, from the earliest unions of the smallest particles to form atoms and molecules to the highest levels of conscious life. The significance of this movement of harmonization—might we say its purpose?—is that it increases the worth of existence, making it more precious and desirable, by intensifying consciousness and enriching it with ever higher values of beauty, understanding, and love. Indeed, I cannot imagine any other significance that the cosmic process might have, for what would be the worth of a Universe of billions of stars thinly scattered through vast stretches of space, with nothing in all this immensity to enjoy existence? All the restless permutations of the cosmic stuff, from the formation of stars to the evolution of life, become meaningful when interpreted as stages in a sustained striving to avoid the utter barrenness of a Universe devoid of creatures that enjoy their existence in it. We are a partial fulfillment of this immensely prolonged effort, its chief beneficiaries in the solar system.

Recognition of our intimate relation to the universal constructive process should make us feel less alone. But intellectual acquiescence in the fact is not enough; we must accustom ourselves to feel, pulsing through our bodies, activating our minds, sustaining us from day to day, the same creative energy that formed the solar system and fashioned all the living beings around us. This vivid sense of our close relationship to all organized beings should make us feel less aloof from them. We are their elder, and perhaps wiser, brothers who must often resist or combat them, but always with the restraint that comes from the thought that they, like ourselves, are products of the process that raises existence to greater awareness and enjoyment, although they appear to have been less fortunate in the outcome.

Understanding our relatedness to the creatures around us, to the whole natural world, should increase our capacity for loving, which is the most effective cure for loneliness. The more widely we can love and, above all, lovingly care for, the less solitary we shall feel amid the immensity of space. While we march at the forefront of the living world, too often lonesome and fearful, with nothing but our own ideals before us, the immense forces behind impel us forward. As we become more sensitive to their direction and intensity, more interested in and careful of the multitudinous creations of these same forces, we shall advance with greater confidence in our future.

24. *The Elusive Queo*

For eighteen months I dwelt beside the Río Buena Vista, a tributary of the Río Térraba on the Pacific slope of southern Costa Rica. My thatched cabin stood close beside the rushing torrent amid small patches of cultivation, bushy pastures, and much resting land covered by dense thickets. On either side of the narrow valley rose steep, forested ridges that swept up to the continental divide in the Cordillera de Talamanca.

With the approach of the winter solstice, as the days became sunnier and drier, I began to hear, issuing from the thickets around me, a bird song of wonderful beauty. It was repeated most frequently in the early morning; often, as I sat at breakfast on my porch, the full, clear notes reached me from the bushes across the grassy roadway. The powerful song was so unlike that of any other bird I knew that I could not even surmise the family relationship of its author. For a long while my efforts to glimpse him were in vain. The dense verdure at the thicket's edge quite concealed the bird who sang so gloriously within it; when I tried to force my way through the tangle of bushes bound together by creepers, I inevitably made so much noise that I drove him away.

It was not until the thickets lost most of their foliage and became more penetrable to vision with the advance of the dry season in February that at last I won a glimpse of the secretive musician. After I learned what to look for, I saw him repeatedly. In appearance he was no less lovely than in voice. He was about 8 inches (20 centimeters) long with a broad, rounded tail. All his upper plumage, including the wings and tail, were slaty black, as were his cheeks. Each side of his forehead was broadly red, which color extended along the sides of his head as a narrowing streak that faded to pinkish above the eye, then continued backward to the hindhead as a thin whitish line. All his central underparts, from chin to tail coverts, were bright rose-red, which on the sides of the breast was invaded by extensions from the dusky upperparts, forming an incomplete collar. The female was similar to her mate, but the rose-red was replaced by tawny.

When, months later, I learned the name of this puzzling bird and could look up its distribution, I found that it has a curiously discontinuous range. It occurs in western Mexico from Sinaloa to Colima, is absent from southern Mexico and nearly all of Central America, reappears on the Pacific side of southern Costa Rica, and extends through Panama to Colombia and Venezuela. The forms occurring to the north and south of the wide gap in its distribution are so different that Ridgway classified them as distinct species, although more recently they have been regarded as geographical races of *Rhodinocichla rosea*.

In the Térraba Valley, to which *Rhodinocichla* appears to be confined in Costa Rica, it is far from common. Years ago, there was a small colony at Rivas in the lower part of the valley of the Río Buena Vista, around 3,000 feet (915 meters) above sea level. I have heard it in the canebrakes along the Río General, somewhat lower in the same drainage system, and have found it at Buenos Aires de Osa, much farther to the east. But at several intermediate points where I have spent months or years I have failed to detect the presence of this bird. I have heard it a few times at Los Cusingos, fewer than ten miles from the point where I first made its acquaintance and a few hundred feet lower, but I could not find a nest here. Yet on this farm and in other localities where I have searched in vain for *Rhodinocichla* are large areas of dense thickets much like those in which I first met it. The factors that control the distribution of this bird are puzzling.

Equally perplexing is its classification. It was first placed in the Furnariidae or ovenbird family, where obviously it does not belong, as it is an undoubted songbird, and since then it has been bandied about among the wrens, the mockingbirds and thrashers, the wood-warblers, and the tanagers. According to the family in which it was placed, it has been variously called thrush-warbler and more recently Rosy Thrush-Tanager. The difficulty is that, although superficially *Rhodinocichla* resembles a mockingbird or even a wren more than it does a wood-warbler or a tanager, it has only nine primary feathers on its wings, like the last-mentioned families, not ten primaries, as in the thrashers and wrens. From my first acquaintance with it, I

Rosy Thrush-Tanager or Queo

tried to learn what affinities were indicated by its habits and also to find an appropriate English name.

As the dry season advanced, I succeeded in glimpsing *Rhodinocichla* more frequently, not only because the thickets where it dwelt were less densely screened by foliage, but because of the loud, rustling sounds it made while searching among the dry dead leaves and other litter that now covered the ground. Sometimes I watched it briefly while it flicked aside the crackling dead foliage with its strong bill in the manner of the Blue-and-White Mockingbird of the Guatemalan highlands, of which, despite its very different coloration and habitat, *Rhodinocichla* strongly reminded me. This seemed to be its chief method of foraging, from which I judged that insects, larvae, worms, and other small creatures that lurk in or beneath the ground litter formed, along with seeds, the bulk of its nourishment. Clark (1913) reported that *Rhodinocichla* eats beetles of at least four species and seeds, especially the gray achenes of a sedge; it also swallows large, irregular grains of sand. But the bird was at all times excessively shy, and if it noticed that it was being watched, even from a considerable distance, it promptly vanished into the depths of the thicket, where it continued industriously to rustle the leaves beyond my view. After the returning rains soaked the ground litter and the limp dead leaves could be stirred silently, *Rhodinocichla* was much harder to find. I marveled that a bird so intensely colored and, to judge by its voice, so numerous in this neighborhood could so consistently elude eyes alert to see it.

Rhodinocichla appears to remain mated through much if not all of the year, and, as in wrens, which likewise maintain pair bonds amid dense vegetation where visibility is narrowly limited, voice seems to be more important than vision in keeping the partners together; hence it is well developed in both sexes and used rather freely. The notes of *Rhodinocichla* are full, mellow, and wonderfully sweet. Its songs are short and varied. Usually they failed to suggest words to me, but once I heard a bird sing distinctly to his mate *Don't you fret, dearie; cheer, cheer, cheerily cheer.* On another occasion one seemed to sing *He gave the merry jump.* Frequently, each song was repeated several times in rapid succession in the manner of mockingbirds and thrashers. Once, while I sat in a blind amid a thicket watching a Variable Seedeater's nest, a male sang one of his lovely verses beyond my sight, while his mate, perching in full view, accompanied him with a melodious liquid refrain that sounded like *witty witty witty witty.* After this outburst of song, both vanished amid the dense vegetation and were not seen again. On another occasion, I watched a pair sing a duet. Although the female's song was much like that of her mate, her voice was not quite so full and strong.

Sometimes I heard these birds utter alternately two different liquid calls, the first of one syllable and the second of two. It was easy to imagine that one member of the pair was calling *gold* while the second answered *silver.* Since I did not succeed in watching the delivery of these notes, I could not exclude the possibility that both sorts were voiced by the same individual.

Often, especially as the long rainy season drew to a close in December, I found *Rhodinocichla* perching near the ground in a dense thicket or canebrake, repeating tirelessly a full, sweet-toned, but rather querulous *queo*. This liquid call, with its variations *querup* and *quero*, was sometimes given in the morning, but I heard it most frequently late in the afternoon, usually when clouds covered the sky. Pleasant as this utterance was, it was sometimes reiterated until I grew tired of hearing it. This mournful call was so characteristic of *Rhodinocichla* that at last it suggested a vernacular name for the bird, and thenceforth I knew it as the Queo (the *que* as in *queen*). This designation, provided by the bird itself, is not only much briefer but seems more appropriate than the hyphenated Thrush-Tanager or Thrush-Warbler, and it has the great advantage that it will still be applicable, however the classification of *Rhodinocichla* is finally settled.

The Queo's breeding season was long. About the middle of February, I found, in the thickets near the Río Buena Vista, a pair accompanied by two juveniles, so recently departed from the nest that the yellow corners of their mouths were still prominent. Less shy than their parents, they were easier to watch, but they neither foraged for themselves nor were given food in my presence. Their upper plumage was browner than on the adults, the under plumage and superciliary stripes paler red; and the dark collar across the breast, which on adults was merely suggested by intrusions of the slate color of the side, was more nearly complete on them.

In the same locality I discovered, in mid April of 1936, the only nest of the Queo that I have seen. It was situated 3 feet (1 meter) above the ground among intertangled bushes and vines in a low, dense thicket. On a foundation of coarse sticks was a shallow, well-made bowl composed of the secondary rachises of the twice-compound leaves of the acacia-like *Calliandra similis*. It contained two white eggs, of which one bore a wreath of blackish scrawls and spots around the thicker end, whereas the other had a few blackish spots scattered at random over the surface, with the exception of the more pointed end.

When I first came upon the nest, both parents approached much closer to me than I had ever seen a Queo before, and in their excitement both sang loudly, one in a voice deeper than the other's, while they perched low in the thicket with their breasts toward me. Through the eggs' somewhat transparent shells, I could see that embryos were just beginning to form. When I revisited the nest two days later, one egg had vanished, and after three more days the nest was empty. I was intensely disappointed by this loss, which extinguished my hope of making detailed studies. Then and in the following year, I vainly searched for another nest in the same locality.

Although anatomists find that the internal structure of the Queo is more closely similar to that of undoubted tanagers than to mockingbirds and thrashers, and recent classifications include it among the tanagers, it differs from them in many ways. Ground foraging with leaf tossing as the chief method of finding food is, to my knowledge, otherwise unknown in the tanager family, although fragmentary observations suggest that it may be habitual in the little-known Chat-Tanager of the mountains of Hispaniola. Tanagers are, with few known exceptions, poor or lacking in song, the females notably so, and I have never heard one duet with her mate; yet both sexes of the Queo sing well and join in duets. Its nest with a foundation of sticks, so similar to that of the Blue-and-White Mockingbird, is unlike that of any undoubted tanager that I know; and incubation by the male, which has been reported of the Queo, has not been confirmed for any unquestioned tanager. On the other hand, the presence of only nine (instead of the more frequent ten) primary feathers of the wings and the red interior of the nestlings' mouths (shown clearly in color photographs by Paul Schwartz) ally it more closely to the tanagers than to the mockingbirds. Only a small minority of the approximately 230 species of tanagers have been well studied as living birds. Perhaps when we know more about them we shall find a few that bridge that gap between the Queo and more typical members of this great family. Meanwhile, we must regard the Queo, like the Donacobius, as a unique bird which does not fit comfortably into any presently recognized avian family.

25. The Growth of Caring

Of all the discoveries that have rewarded me during more than half a century of studying birds, none gave me greater pleasure at the moment when it was made or more substance for contemplation in subsequent years than that of the young Groove-billed Ani who helped his parents attend the nestlings of their second brood. Anis are slender black birds of the cuckoo family whose long tails seem so loosely attached that they are in some danger of falling off. They have high-arched, narrow black bills, bare black faces, and distinctive voices high in pitch. Widespread in tropical America, they are a familiar sight in pastures, where they forage close beside the heads of grazing animals and with clumsy agility capture grasshoppers and other insects stirred up by the horses and cows. The domestic arrangements of these birds are as strange as their appearance. Two or three pairs may join forces to build a single nest of coarse sticks lined with freshly plucked green leaves, upon which they lay chalky white eggs in a common heap, generally about four to each female. All the members of the communal group, males as well as females, take turns incubating them, and later all share the task of feeding the ugly naked nestlings, whose development and feathering are surprisingly rapid.

The young ani whose behavior so delighted me was the single surviving offspring of a pair that nested alone. At their first nest in an orange tree, I tried to make one member of this pair touch a small paintbrush that I stuck into the nest, for without some artificial means of recognition I could not distinguish the male from the female. I was successful beyond expectation when one of the anis carelessly rubbed against the brush and smeared its face liberally with white paint. I called this bird Whiteface, its mate Blackface. Until I watched the latter lay an egg, I was uncertain of their sexes. Whiteface, the male, was the more zealous parent, sitting in the nest throughout the night (the first male bird that I found performing this usually feminine office, although the habit is widespread among male woodpeckers), bringing the larger share of food to the young, and defending them more bravely than Blackface did. Sometimes he buffeted the back of my head when I visited the nest.

Now Whiteface and his mate had hatched a second brood in a lemon tree, 8 yards (7 meters) from the orange tree that had held their first nest. While they incubated, the young ani of the first brood, nearly full grown but still without grooves on his bill, had often rested on the nest's rim beside his sitting father or mother. At times he carried an insect or small lizard to the nest and might have given it to his incubating parent if the latter had accepted it. After three nestlings hatched, the juvenile ani, then slightly over two months old, helped feed them. In four and a quarter hours, he brought food eight times, while Whiteface fed the nestlings twenty-nine and Blackface fourteen times. He zealously defended his younger brothers and sisters, venturing close when I visited them and protesting with an angry *grrr-rr-rr*. In the absence of his parents, he tried to defend the nest alone. Already he showed more spirit in guarding it than Blackface, his mother. From this early ardor in defending the brood, I surmised that he was a male, for his father was so much bolder than his mother when the nest appeared to be endangered. How direct and unforced was his education, the natural outcome of his continuing intimate association with his parents! In helping them so spontaneously, he was preparing himself for effective parenthood in the coming years. With all our verbose theorizing about education, we often fall lamentably short of the simple precepts of nature.

In subsequent years, I watched young birds of several other species help feed and guard the nestlings of later broods, and a growing number of examples of this behavior are reported in the literature of ornithology. Often the youngsters begin this precocious parental activity in the season when they hatched. If they do not pair until the second or third year after their birth and are of sociable species, they may attend nestlings not their own when they are a year or more of age. Juvenile helpers only a month or two old are found among bluebirds, swallows, and gallinules. Less often they have been noticed among Northern Cardinals, Southern House-Wrens, and Golden-masked Tanagers. Helpers at least approaching one year of age but still apparently adolescent and not ready to breed occur among Brown Jays, Banded-backed Wrens, and the small toucans known as Collared Araçaris, all of which inhabit southern Mexico and Central America. Other examples are mentioned in chapter 5.

The inclination of young birds to help their parents feed subsequent broods is apparently more widespread than its manifestation. In many species, the young are not only thrown upon their own resources but are driven from the parental domain as soon as they can forage well enough to support themselves. Accordingly, they enjoy slight opportunity for contact with their younger siblings, which if they saw for a slightly longer period or saw at all, they might well feed and protect. When we remember the many perils that threaten eggs and nestlings, the host of hungry mouths, reptilian, mammalian, and avian, ready to devour them as soon as they find them, we recognize a certain advantage in sending away the older offspring. For the greater the number of attendants, the more movement and bustle in the nest's vicinity, the more likely it is to attract hostile eyes. This is especially true if some of the attendants are inexperienced young, who may be less cautious than the parents themselves.

In the house-wren of tropical America, which like its North American counterpart breeds in a variety of holes and crannies, juveniles may attend nestlings of the following brood if the parents permit them to sleep in the nest space after this brood has hatched. But only exceptionally are fledged young allowed to make their home for so many weeks in their natal box or gourd or cavity in a tree, with the result that one rarely finds young helpers among Southern House-Wrens. It is a pity that the need to preserve secrecy at the nest, to refrain from all unnecessary activity in its vicinity, along with the territorial exclusiveness widespread among birds have led to the habit of holding older offspring aloof and thereby reducing their opportunities to be helpful. Were it not for this, an activity charming to watch which increases the young helpers' safety while it educates them would probably be much more frequent than we find it. But it is not only among birds that a hostile environment prevents life from attaining the most beautiful expressions of which it is capable.

Among the species in which juvenile helpers are of infrequent occurrence are the exquisite little Golden-masked, or Golden-hooded, Tanagers of Central America. These birds of the treetops have a long breeding season in which they may raise two or rarely three broods. Although they are, like most tanagers, monogamous and live in pairs throughout the year, they are not strongly territorial, and I have thrice found nests to which three or four adults were bringing food. When the young of an early brood linger near a later nest, the parents are sometimes antagonistic. But their efforts to drive older offspring to a distance are at most mild and often ineffectual, and occasionally their indulgence is rewarded by the acquisition of an assistant.

Some years ago, a pair of Golden-masked Tanagers nested in the Calabash trees in front of the house. In May they reared a single fledgling who took wing at the usual age of fourteen days. Two days later, they started a new nest about 30 feet (9 meters) from the first. By mid June, when they were feeding nestlings, they were helped by a young tanager, just beginning to molt from the dull green juvenal plumage to the black, white, blue, turquoise, and gold of the splendid adult attire of both sexes. If, as was probable, it was the juvenile from the neighboring nest of this pair, it was only forty-five days old when I first saw it bring food to its younger siblings. Often it went, with the parents or alone, to the nearby feeder, where it ate bananas, then carried a billful to the nestlings. It also brought food of other kinds from more distant sources. When I first watched, the youngster fed the nestlings about as often as either of the parents, although it usually brought smaller portions. As the days passed, it became less diligent, bringing food no more than three times in an hour, in some hours giving the nestlings nothing, although it might follow the parents when they came together with food, as they did throughout the day, from four to nine times per hour.

The behavior of this young Golden-masked Tanager reminded me of that of human children when they first engage in grown-up activities. How eager they are to begin, how important they feel when permitted to help their elders—and how soon, when young muscles tire and the task proves to be not so easy as it appeared, do their efforts relax!

In a deeper sense, the precocious participation in parental activities by this and other juvenile birds that I have watched revived almost forgotten memories of childhood. The human infant, born while still so weak and helpless that it seems almost embryonic when compared with many other newborn mammals, requiring so exacting a regimen of bottles, baths, diapers, pins, and powders, seems too delicate to be entrusted to an older brother or sister not yet ten. Instead, they have pets, a dog or cat, a hen or pigeon, perhaps some wild foundling from the woods or fields, or, if not a living creature, an effigy in human or animal form, a doll or quadruped or bird, made of porcelain or stuffed with cotton. And the cherished object, living or inanimate, elicits ineffable love and devotion from the young heart of its owner. How fondly he caresses it! With what pride he shows it to his playmates! He imagines himself capable of the highest feats of courage in its defense. He is desolate if it is lost, broken, or dies.

The human child's attachment to his living pet or lifeless effigy has much in common with that of the young bird who feeds, defends, and occasionally broods nestlings. Such cherishing behavior is also seen in other young animals. I have watched yearling heifers lick and fondle little calves as though they were mothers. Juvenile Chimpanzees and other nonhuman primates eagerly embrace and groom infants of their own or other parents, and the young of jackals and other canines feed and play with their mother's cubs. In both mammals and birds, some of the impulses that make good parents arise long before parenthood is physically possible.

In its inconstancy, too, the child's treatment of his pet often resembles the juvenile bird's wavering attentiveness to its younger siblings. Just as the immature Golden-masked Tanager was most irregular in bringing food to the nestlings, who needed to be fed at least several times each hour, so a careless boy one day nearly kills his pet with excessive attentions, the next day half starves it by neglect. In both birds and men, steadfast attentiveness matures slowly with the years.

The young child's capacity for loving devotion to things dependent upon itself is one of its most precious psychic assets. It is obvious how, when properly matured, this capacity will make good wives and husbands, good parents and dependable homekeepers. Moreover, it is interesting to speculate how much of our art, literature, and science has here its deepest roots. Are not much of our best poetry and painting attempts to give permanence, by words or visual images, to the fleeting things that we love and would cherish? Some scientific work may be motivated by curiosity alone, the effort to answer the eternal how and why, but probably much more springs from deep affection for the objects of study. To produce scientific work of the highest worth, the botanist must love her plants, the ornithologist his birds, the geologist the Earth with all the stony tablets whereon its history is inscribed, and the astronomer must feel a certain affection for the distant stars, make them her protégés and believe that they somehow depend upon her, if not for the continued emission of their light and maintenance of their fixed positions in the firmament, at least for their interpretation and understanding among mankind.

As we have seen, parental impulses, which prompt an animal to cherish and protect young of its own kind or substitutes for them, appear in both birds and mammals long before the maturation of sex. A wholesome education should foster these impulses at an early age. At first they should be focused upon inanimate objects: dolls, toys in the form of animals, or the like. These may be tossed about, alternately caressed and neglected, without causing pain other than to the child who will be forlorn when the loved object is broken or lost. His distress will probably be short-lived, and through it he may learn that things must be treated gently and kept in their proper places if they are to endure. Later, the child may have a living pet, and its selection calls for deep consideration. Animals that have not been domesticated are unsatisfactory; if deprived of their freedom, they either escape or languish and die in captivity.

They may use teeth or claws too strenuously. Perhaps a puppy is the best pet for a child; the young of men and dogs have romped and tussled together for so many generations that each can well take what the other gives. A colt or pony is the ideal companion for a growing boy. Our civilization would be more wholesome if more lads grew up where ownership of such an animal were practicable.

Possession of a doll or pet or anything dependent upon us can, under adroit adult supervision, lead unobtrusively to the most valuable lessons to hands and head and heart. The girl can make clothes for her doll, the boy a kennel for his dog or a dovecote for his pigeons. A living animal's need of food and attention at regular intervals promotes the development of orderly habits and responsibility. The child must learn consideration for the feelings of the things dependent upon him—not to play too roughly with his dog or ride his pony too long and hard. Since at a certain stage of intellectual development earnest young people sometimes entertain the delusion—which is not wholly a delusion—that they can make a better world, there is no limit to the heights to which the impulse to cherish and protect may lead.

The cherishing temperament is best developed in the ambience where it most naturally flourishes—the home. The ideal home for a child is one that gives him duties and responsibilities and fosters the feeling that he contributes importantly to the welfare of the household and all its inhabitants. The sooner he can participate in the family councils, learn the reasons governing important decisions and the wisdom of expenditures, learn also prudence in speaking about family matters with outsiders, the better for him. Too many modern dwellings, where nearly everything necessary comes through wires and pipes or in delivery vans, where the chief problem of homekeeping is how to make, at a distance, enough money to pay for it all, are poor homes for a child. One of the wealthiest men in America lamented the difficulty of finding for his sons, in the modern urban house, "some substitute for the old-fashioned wood-pile."

In the home, and especially on a family farm, children have the benefit of the adult companionship that is no less necessary for them than playmates of their own age. The excessive herding together of children is neither natural nor wholesome. Among mammals and birds with a well-knit social life, forming cooperative societies rather than formless, unintegrated flocks and herds, the young are rarely at any stage of development segregated from the adults. Association with adults tends to make the child adult-minded and responsible; with only children as intimate companions, he may remain child-minded and irresponsible. It is natural for a boy or girl to want to help in the tasks of father or mother or an employee; they can learn much by doing so, often so much more painlessly than in school that the maximum opportunity for such participation should be provided. Sensitive, timid children, who shrink from the rough physical contacts of the gymnasium or playing field and hence are commonly relegated to the less important positions, where much of the time they stand and watch more robust playmates take the exercise of which they themselves have greater need, often join with a will in the labors of home or farm, from which young bodies reap immense benefit while characters are formed. From such early participation in the tasks of home and family, the impulse to care for and to cherish, innate in the young child, is strengthened and disciplined until it expands far beyond childhood's narrow horizon.

It is useful to distinguish between "caring for" and "caring about." "Caring for" denotes the overt activity, without reference to whatever feelings may accompany it. "Caring about" implies deep concern or cherishing sentiment for the object of care. We may be obliged to care for things that we care little about, and, at the opposite extreme, we may care intensely about things that we can hardly care for, as when we are distressed to learn that tropical forests which we may never see are relentlessly destroyed. As, in the evolutionary sequence, psychic life intensifies, caring for may blossom into caring about. In the individual, caring about may lead to caring for, as when we discover how to advance a cause for which we have long been concerned.

With the exception of birds and mammals, most animals care for nothing except their individual selves; they drop their eggs in the water or deposit them on vegetation, other animals, or some lifeless substrate, then neglect them. Even if eggs are retained in the body until they hatch, as in a number of fish, amphibians, and reptiles and not a few invertebrates, the newborn young commonly receive no parental attention. However, among both insects and "cold-blooded" vertebrates, many exceptions to this rule are known. Wasps, ants, bees, termites, and a few other insects both solitary and social make more or less elaborate preparations for the reception of their eggs, and they may also diligently care for the larvae that hatch from them. Many social insects vigorously defend their nests or hives. Obviously, they care for their nests and progeny, but they are so different from us that we can hardly surmise whether they are stirred by cherishing feelings—whether they care about them.

With the notable exception of social insects, among which workers unable to reproduce attend the callow brood while sterile soldiers defend them, invertebrates and "cold-blooded" vertebrates rarely attend the offspring of other individuals. Among birds and mammals, the capacity for caring expands. In many species of birds, juveniles and young adults attend nestlings and fledglings that they did not beget, usually but not always their younger siblings. Moreover, breeding adults, sometimes parents who have lost their broods, may diligently feed the young of other pairs, of their own or quite different species in the most diverse combinations. Caring for young that may differ in size, color, voice, and other features from those of the altruist's own species is an activity that appears to escape strict control by the genes and implies an ability to recognize fundamental similarities despite differences in detail. This contrasts with the behavior of animals mentally less advanced, such as cichlid fishes, among which chemical stimuli, or taste, finally determine whether parents will protect or swallow foreign young introduced among their own fry (Noakes and Barlow 1973).

In chapter 27 I give reasons for believing that parent birds not only care for but likewise care about their progeny or are emotionally attached to them. Young helpers, too, often appear to feel strongly about the nestlings they feed and protect. The juvenile Groove-billed Ani was obviously perturbed when I examined the nestlings that he fed and guarded. Long ago, at the edge of a Guatemalan banana plantation, I studied a Brown Jay's nest attended by five grown birds, parents with three nonbreeding helpers. One of the latter, whom I called Pied-bill, surpassed the other four in zeal as the nest's protector. Whenever I climbed the willow tree to view the nestlings, Pied-bill worked itself into a frenzy, screaming loudly and darting to within a few inches of my head. Lacking the courage to attack the intruder, this young jay of undetermined sex pecked the branch where it perched and tore to shreds a nearby banana leaf. After the other four jays had exhausted their protests and retired to a distance, Pied-bill remained close to me, noisily expressing its disapproval, until I descended to the ground. After such a demonstration, I could hardly doubt that this bird cared feelingly about the nestlings in the willow tree.

As a measure of evolutionary advance, I would place the capacity to care above morphological or physiological features and perhaps even above intelligence. From its roots far down in the animal kingdom, we trace the growth of this capacity through those relatively few fish, amphibians, and reptiles that devote more or less attention to their immediate progeny, to mammals and birds, many of which care for young not their own, and finally to its flowering in man. Appearing early in the human child, who cares lovingly for dolls or pets, it grows with expanding outlook until the most spiritually advanced humans not only care devotedly for their homes, children, and much besides but care intensely about matters too great for individuals to care for, such as the future of mankind and the health of their planet with all its living cargo. More than man's dominant position among the animals, more than all his intellectual and material achievements, his divine capacity to care deeply and widely entitles him to be considered the most advanced product of evolution.

26. *The Smallest Tanagers*

On a morning in late March of 1932, in the foothills of the Sierra de Merendón at the edge of the Motagua Valley in Guatemala, I noticed a very small black-and-yellow bird with a dry grass blade in his bill. Following him, I found his half-finished nest in a cranny at the top of a decaying fence post beside a corral amid pastures with scattered trees. The nest was so well hidden behind a small aroid growing on the post that I would not have discovered it if the bird had not led me to it. A roofed structure with a round doorway in the side, it was composed of fine tendrils and other bits of vegetation, all bound together with cobwebs. The outer shell appeared to have been completed, for the black-and-yellow male and his olive-green mate were lining it with small, dry grass blades. Nearly always they arrived together, each with a fragment of grass in its bill, and one promptly vanished behind the arrow-shaped leaves of the aroid. When it emerged, the other entered the nest to place its contribution. Each always arranged with its own bill whatever it brought instead of passing it to the partner already inside. Either might enter first, and each waited nearby while the other worked within. The waiting male repeated a fine, metallic *pink pink* that reminded me of the notes of a Northern Cardinal as he goes to roost on a chilly winter evening. The moment the second bird reappeared, the pair of Yellow-throated Euphonias flew off together to seek more material.

Such was my introduction to the euphonias, a genus (*Euphonia*) of about twenty-six species of small to very small birds widespread on the mainland and islands of tropical America, from Mexico to northern Argentina. Although with their close relatives the chlorophonias they are included in the tanager family, they differ in many ways from all the other tanagers that I know. They are stout little birds with short and rather thick bills. The males are usually clad in black glossed with purple or steel blue, with bright yellow on the forehead and often more or less on the crown and yellow under-

Tawny-bellied, or Spotted-crowned, Euphonia male (above) and female perched on Topobea durandiana, an epiphytic shrub of the melastome family

parts, with or without a black throat. The much plainer females are commonly olive-green above and yellowish below. However, other color patterns are found in both sexes. Male and female of the widespread Blue-hooded Euphonia have blue crowns and napes. The male Golden-sided Euphonia of northeastern South America is nearly everywhere glossy purplish black, with yellow tufts on the sides of his breast. The male Olive-backed Euphonia of the Caribbean lowlands of Middle America is largely olive-green, with a tawny abdomen, not greatly different from his mate.

Euphonias travel singly, in pairs, or in small flocks, often with other frugivorous birds. Although some species are most often seen in forests and others in open country, like many birds that wander widely in search of ripening fruits, most appear not to be restricted to a single habitat but frequent both woodland and clearings with scattered trees. Euphonias are most often seen at forest edges or in open groves, shaded pastures, parks and gardens, roadside trees, and scrubby growth. They prefer the crowns of trees to undergrowth. Most abundant in warm lowlands, a few range upward in the subtropical zone or more rarely the temperate zone. In Guatemala I found Blue-hooded Euphonias up to nearly 10,000 feet (3,050 meters). At high altitudes euphonias are found much less frequently than their green relatives, the chlorophonias.

Largely frugivorous, euphonias eat a variety of small fruits, including berries of melastomes and epiphytic aroids, bromeliads and cacti, small figs, and the green fruiting spikes of Cecropia trees. They are among the few birds that have been observed eating the spikes of the pipers so abundant in the more humid regions of tropical America. Although less eager for small seeds enclosed in soft, oily arils than other tanagers are, they do occasionally take them, including those of the tree *Protium* and an epiphytic vine of the marcgravia family, *Souroubea guianensis*. Berries of the parasitic mistletoes of the loranthus family attract them strongly and appear to be the preferred food of some species. After partly digesting the berries, the euphonias void the seeds in their droppings still covered by a very sticky mucilage which attaches them to the branches and twigs on which they fall and germinate. Although by no means the only eaters and distributors of mistletoe seeds, euphonias are apparently their principal disseminators in tropical America. Indeed, mistletoes and euphonias appear to have evolved together because each is so useful to the other. Whatever small fruits they eat, euphonias remove the skin before they swallow the flesh. From the ground beneath a tree where a number of Blue-hooded Euphonias were feasting on mistletoe berries, I picked up many parchmentlike husks from which the entire contents had been neatly removed.

One morning I watched a Streaked Saltator digging into a full-grown but still green guava in the top of a tree behind our house. It pecked out small bits of the fruit and chewed them, dropping fragments from its bill. A male Tawny-bellied Euphonia, waiting for his turn at the fruit, hopped around and gathered from the leaves pieces that the larger bird had dropped. Becoming impatient, he clung to the other side of the saltator's fruit and started to eat, but was driven off by a peck. When he tried again to share the guava, the saltator did not repulse him. After they had eaten together for a while, the saltator flew away and the euphonia had it to himself. He did not enjoy it long, because the saltator's mate, who had been waiting nearby, now claimed it. After the second saltator had satisfied itself, the euphonia returned to the fruit and ate freely, pecking out bits from the inside and mandibulating them, just as the thick-billed saltator did. Before the tiny bird had finished, a Blue Tanager took the fruit from him. The larger tanager ate little, and when it left the euphonia resumed his meal. Soon, however, he was displaced by a Buff-throated Saltator and flew away. The euphonia had spent much more time waiting than eating. Euphonias prefer small berries to larger fruits. Of all the tanagers that frequent our garden, they are least attracted by bananas and seldom visit the feeder.

In a riverside thicket one morning in December, I watched a pair of Tawny-bellied Euphonias sip nectar of the slender, acacialike shrub *Calliandra similis*, which was flowering profusely. The numerous long, slender red stamens sprang in a cluster from the little, five-toothed green calyx; there was no corolla. As the rays of the rising sun pierced brightly into the thicket, these tiny green cups brimmed with sweetish liquid, which if not pure nectar was nectar diluted by the past night's rain. Even the short bills of the two euphonias could reach the liquid. They appeared to squeeze the calyxes with their bills to press out nectar that they could not reach with their tongues. With them was a Golden-masked Tanager, also enjoying the sweet liquid.

While some euphonias subsist mainly upon fruits, others, including the Trinidad and the White-vented, spend much time searching for insects. In Trinidad, the Snows (1971) saw the former gleaning insects much as I have watched the latter engaged in the same activity in Costa Rica. These euphonias hunt on slender twigs, often thinner than a pencil, mostly high in exposed treetops, along which they progress either by hopping, by sidling along for an inch or so, or by about-facing, alternately to the right and the left, with each reversal of direction placing the foot that was behind ahead of the other foot. As they advance, they bend over to examine from opposite sides the lower surfaces of the twigs; but what they find there is nearly always too small to be seen from the ground. This manner of foraging is widespread among tanagers, especially the lovely little species of *Tangara*, who prefer thicker, horizontal branches well covered with mosses and lichens. Euphonias also pluck larvae and mature insects from living foliage and rummage among dead leaves caught in vine tangles well above the ground.

Male euphonias feed their mates, and occasionally the female gives food to her companion, as I have seen both Tawny-bellied and White-vented euphonias do, the latter at the nest, the former while they hunted through the trees. Likewise, a male Olive-backed Euphonia regurgitated food to his partner.

In a family of birds more renowned for beautiful plumage than for song, euphonias are exceptionally songful, but as songsters they win admiration for the duration rather than the quality of their performances. Their songs, often pleasant but perhaps never outstandingly lovely, are usually medleys of whistles and soft, liquid notes, frequently in doublets or triplets, intermingled with various dry, harsh, or throaty notes. Their manner of singing is not unlike that of siskins. The performance of the Blue-hooded Euphonia, one of the better songsters, is a prolonged, rambling sequence of weak, tinkling notes, now high, now low, tumbling over each other without set phrasing—an artless outpouring yet delightfully cheery as it drifts down from a high treetop where a flock of mixed ages and sexes is eating mistletoe berries. Its wild disharmony reminded me of the Bobolink's rhapsody, although it was less powerful and impassioned.

Just as the euphonias' songs tend to be long continued, so, too, are their complaints when disturbed at their nests. Often, when I have examined a nest, the female has perched nearby, continuously scolding with a sharp *chip* or a throaty *churr* as long as I remained. A female Tawny-bellied Euphonia whose nest had just been raided by a squirrel continued for at least a quarter of an hour, with scarcely a pause, to protest in this manner, while she flipped out her wings simultaneously and turned from side to side. Finally, she ate the greenish fruits of a shrub of the melastome family that grew on the nest tree until her mate arrived and she flew away with him. Another female, whose nest hung above a pool in a mountain torrent where I frequently bathed, would complain with throaty rattles followed by high, chaffy notes the whole time I was in the water.

Surprisingly, at least two species of euphonias are excellent mimics. As one would expect from the character of their own songs, they tend to imitate the call and alarm notes of their neighbors rather than the melodies of the best songsters. In Trinidad, Violaceous Euphonias interspersed their rambling songs with the calls of seventeen species in ten families, ranging from parrots and hummingbirds to thrushes and other tanagers. In Amazonia, Thick-billed Euphonias imitated twenty-five species in thirteen families, including hawks, toucans, woodpeckers, flycatchers, and Troupials. In Panama, females of this species, driven from their eggs, mimicked the alarm notes of other birds nesting in the vicinity so well that one incited a Yellow-green Vireo to scold the intruder (Snow 1974; Remsen 1976; Morton 1976).

Tanagers and other birds that build open nests nearly always roost amid foliage, but those that breed in covered structures frequently sleep in them or in similar shelters. Accordingly, I would expect to find all euphonias sleeping in crannies of some sort, but I have discovered the dormitories of only the Tawny-bellied Euphonia. For many years, until most grew old and died, several Calabash trees stood at the top of the high bank in front of our house. The rough bark of these small trees with long, stiff branches was covered with masses of moss and green and brown liverworts, among which grew orchids, bromeliads, and other larger epiphytes. For twelve years, euphonias slept in these trees, sometimes a solitary individual, sometimes a male and a female a few feet apart, and for long intervals two males and a female in the same tree. Always they lodged singly, usually in snug niches in the thick cushions of brown liverworts, with only the yellow breast of a male or the tawny breast of a female visible in the narrow opening in the side of a cushion. Often, too, they slept amid the closely clustered, slender stems of a small orchid growing on the tree beneath a protecting mass of liverworts. Although usually the sleeping euphonias were at least partly exposed, one evening a female, after repeating dry, chaffy notes for many minutes, dived into a dangling mass of liverworts and lichens about a foot long where she was completely invisible.

The root of an epiphyte or the rhizome of a fern, stretched an inch or so beneath a horizontal branch covered with brown masses of the liverwort *Frullania*, offered a sheltered retreat for a male euphonia. The narrow space between the rhizome and the branch seemed difficult to enter, but the bird shot into it so swiftly and deftly that I could not see how he did it. With his head turned back and his ventral plumage all fluffed out, he looked, in my flashlight's beam, like a little ball of bright yellow feathers. He grasped his perch with one foot; the other was drawn up and hidden in the yellow mass. His back was pressed against the branch above him at a point where a luxuriant growth of the liverwort stood out broadly on both sides, forming a wide, spongy brown roof above him on wet October nights.

For nearly two years, with occasional absences, a female slept in a large tuft of the brown liverworts hanging beneath a horizontal limb 10 feet (3 meters) above the lawn. Sometimes her absences were inexplicable, but at other times they were caused by a Bananaquit. These diminutive, yellow-breasted birds are indefatigable builders of cozy covered nests, both for their eggs and young and for dormitories in which they sleep, always singly, throughout the year. Occasionally, however, evening finds them without a roof of their own, and then they try to enter the covered nest of some other small bird, often a flycatcher. One day I noticed that a Bananaquit had arranged fresh straws around the doorway of the pocket amid the liverworts where the euphonia had already lodged for many months, reducing the orifice. When the euphonia arrived at her dormitory, she paused, clinging in front, repeating her queer *churr*. When she tried to enter, she was greeted by pecks from the small intruder. Nevertheless, she persisted in pushing in until she and the Bananaquit tumbled out, locked together, and fell to the ground. Separating, they flew up to the pocket at the same time, and again they clutched and dropped to the lawn. Finally, the euphonia returned alone and remained for the night. Afterward, the Bananaquit added so much material to the front of the pocket, making the entrance face downward instead of sideward, that the euphonia relinquished it to the invader.

Once, in early August, I watched a female Tawny-bellied Euphonia cling to a tuft of mosses and liverworts hanging from a high branch while she tore out pieces that she dropped, making a pocket. As far as I could learn, this was never used for either sleeping or nesting, but her activity suggested that these euphonias may prepare the niches amid epiphytes where they sleep or build their nests. The pockets in the Calabash trees where they lodged were never lined. Sometimes the euphonias retired silently, but on certain evenings the males continued for many minutes to pour forth an almost continuous stream of dry, rattling, and, less often, liquid notes before they darted into their niches. In the dim light of dawn, they shot silently out and away.

As far as is known, all euphonias build closed nests with a side entrance as do their relatives, the chlorophonias and the Jamaican Euphonia. For these roughly globular structures, they seek a nook or cranny in the most varied situations. Forest-dwelling species, such as the Olive-backed and Tawny-bellied euphonias, often choose a pocket amid the mosses, liverworts, lichens, and larger epiphytes that thickly cover upright trunks in humid woodland, similar to the situations in which the latter often sleeps. A stout segment of a dead branch, well covered with such aerial growths, that has broken away from the base of the limb and hangs below it, attached only by the root of an epiphyte, offers a swaying site to Tawny-bellied Euphonias. These nests are often screened by ferns, bromeliads, orchids, or other plants growing amid the moss. Dangling masses of epiphytic roots are chosen by Tawny-bellied and White-vented euphonias. In the mountains of Mexico, Blue-hooded Euphonias hid their nests in long, pendent gray tufts of "Spanish Moss" (the bromeliad *Tillandsia usneoides*) (Robins and Heed 1951).

Euphonias that prefer open country select different sites. A cranny in the top of a decaying fence post, or a site amid the shoots that sprout thickly around the truncated top of a living post, is often chosen. Shallow cavities in trees, high or low, especially if screened by epiphytes, are sometimes occupied. A nook amid the matted roots of orchids, growing on posts or in a wire basket hanging near a house, is an attractive location. In lieu of a more sheltered site, euphonias build between close-set, upright twigs of a tree. A pair of Yellow-crowned Euphonias nested in an unusual situation: in a large, curled dead leaf of a Cecropia tree that had lodged in the crown of a banana plant, close beside the bunch of green fruit. Violaceous Euphonias frequently place their nests at the tops of banks beside paths, and Yellow-throated Euphonias sometimes nest in banks above streams or roadways in holes that they probably did not dig themselves (Skutch 1954, 1972; ffrench 1973).

Occasionally, euphonias occupy a nest built by some other animal. In Venezuela, I watched a pair of Thick-billed Euphonias carry material into the bulky domed nest of a Kiskadee Flycatcher, a bird much larger than themselves. This borrowed structure rested upon the top of the much bigger, many-chambered nest of interlaced twigs that a pair of Rufous-fronted Thornbirds had hung from a mango tree above the edge of a small pond. In Belize, a Yellow-throated Euphonia occupied the long, slender nest, with a shallow niche in its midst, that Royal Flycatchers had suspended from a liana over a forest stream (Russell 1964). Even an insect nest may be chosen by the euphonias. In Costa Rica, a pair of Yellow-crowned Euphonias built between the layers of brood cells of an old, dilapidated wasps' nest in a quite exposed situation about 100 feet (30 meters) up on a dead branch of a fig tree. As far as I could learn, the birds never incubated there. From the foregoing examples, it is evident that euphonias nest at any height from the top of a tall tree to near the ground or even beneath it, but in my experience those that live in forests tend to build at greater heights than those that roam through open country.

Male Yellow-crowned, Yellow-throated, and Thick-billed euphonias and Golden-browed Chlorophonias apparently need more than a year to acquire full adult coloration, as they not infrequently mate and nest in a plumage intermediate between that of females and juvenile males and that of mature males. Sometimes these young males are difficult to distinguish from their partners. Pairs of Yellow-crowned Euphonias, and doubtless other species, select the nest site together, sitting alternately in promising spots to try them, sometimes with material in their bills.

In addition to the Yellow-throated Euphonias mentioned at the beginning of this chapter, I have watched nest building by one or several pairs of six species: Yellow-crowned, Tawny-bellied, White-vented, and Olive-backed in Costa Rica; Purple-throated in Peru; and Thick-billed in Venezuela. In every case, both partners shared the task. If they did not contribute equally, either the male or the female might work harder. The forest dwellers gather their materials from epiphytes on trunks and boughs rather than from the ground. At all the nests that I watched, whatever the species, the two regularly came and went together, and each always placed in the nest what it had brought instead of passing it to the co-worker already inside. At some nests either sex might enter first, whereas at others this was usually or nearly always the male—an order of precedence that would be more consistently observed while feeding nestlings. After arranging its contribution in the nest, the first bird always waited nearby until the second had done the same. Sometimes one or both would enter again, always singly, with an empty bill or with a bit of material that it had gathered from a neighboring bough while it awaited its partner. Then the two would fly away together to search for more. The mates were inseparable; if one had not found anything suitable when the other was ready to return to the nest, it would accompany its partner with an empty bill.

The completed nest is a cozy globe entered through a narrow doorway in the side which is sometimes shielded by a visorlike projection. Nests built in mossy situations in the forest usually contain much green moss, fine richly branched roots of epiphytes, perhaps diminutive filmy ferns or small plants of the epiphytic fern *Rhipidopteris peltata*. Nests in open situations consist of dead leaves, fine stems and petioles, rootlets, fibers of various sorts, seed down or cotton if available, and often green moss. Whatever the site, cobwebs are used to bind the materials together. The lining may be composed of dry grass blades and strips of broader leaves, flakes of bark, fibrous rootlets, or shiny black strands of the fungous growth often called "vegetable horsehair."

Euphonias usually lay three or four eggs, sometimes only two, more seldom, as in nests of the Yellow-throated Euphonia, as many as five. They are as a rule deposited early in the morning, usually on consecutive days. The tiny white or whitish eggs are speckled or blotched with shades of brown, varying from reddish to chocolate, most heavily on the thicker end but sometimes rather uniformly over the whole surface. Occasionally the eggs bear a few small, irregular black marks.

Only the female incubates, sitting in her well-enclosed nest looking outward. Frequently an incubating euphonia has clung to her eggs while I looked in at her with my eyes less than a yard from hers or until my hand nearly touched her. Finally darting out, she has flitted around nearby, complaining with a rapid flow of sharp or throaty notes as long as I stayed near her nest.

In six hours of a morning, a Tawny-bellied Euphonia took five sessions on her eggs, ranging from 16 to 80 minutes and averaging 54.4 minutes. Her five absences ranged from 13 to 29 minutes and averaged 19.8 minutes. She covered her eggs for 73.3 percent of the six hours. Another Tawny-bellied Euphonia, watched morning and afternoon, sat more patiently. Her five sessions lasted from 77 to 108 minutes, with an average of 86.8 minutes; her five recesses from 13 to 43 minutes, with an average of 29.8 minutes. She incubated for 74.4 percent of ten hours. A White-vented Euphonia, watched all day, sat for sixteen intervals of 12 to 53 minutes, with an average of 30.2 minutes. Her seventeen recesses lasted from 4 to 26 minutes and averaged 12.5 minutes. She was in her nest for 69.3 percent of a day of twelve and a half hours, during which her mate twice regurgitated food to her after she had entered. When he offered to feed her in front of the nest, she refused. Once, as she arrived, he begged with open mouth and she responded by regurgitating several portions of food to him. Another female White-vented Euphonia, watched all of one morning and all afternoon of another day, incubated for 65.1 percent of twelve and a half hours without receiving food while in view. A female Yellow-throated Euphonia covered her five eggs for 68.8 percent of a morning, with sessions of 32, 54, 69, and 83 minutes and absences of 16, 8, 44, and 40 minutes.

When a female euphonia leaves a high nest, whether spontaneously or because of disturbance, she drops from her doorway like a dead weight and falls almost to the ground before she curves upward and flies into the trees. This mode of departure, frequent among small birds with high nests, might mislead a predator which has the nest in its field of vision but has not yet discovered it. The departing bird's flight often appears to start from the point where she turns upward; here, near the ground, is where one who had not noticed the abrupt fall might search for her nest.

Returning to her nest after a recess, the female is often accompanied by her mate. As she flies to her doorway from a nearby perch, he escorts her so closely that he appears to be racing to reach it first. She always wins and shoots deftly through the narrow doorway while just before reaching the nest he veers aside and flies away. Such a spectacular return to the nest is widespread among birds of which only the female incubates in a covered structure entered from the side, including not only the four species of euphonias and one of chlorophonias that I have watched while they incubated but tody-flycatchers and such larger birds as Black-faced Grosbeaks and Bright-rumped Attilas. If such dashing escort is more than a pleasant courtesy, it may be a ruse for deceiving predators. To enter a blind in front of a nest, the watcher sometimes arranges to be accompanied by a second person, who ostentatiously walks away after the watcher is hidden within. This is done to deceive the parent bird, who supposedly cannot distinguish between one and two and accordingly behaves as though the blind were empty. If predators can be similarly duped, the visible departure of one bird after the arrival of two might make it appear that none had entered a nest.

At one nest of the Yellow-crowned Euphonia, the incubation period was about fourteen days, but I suspect some irregularity here, because at another nest of this species it was sixteen days. However, the incubation period of euphonias appears to be variable, for at one nest of the White-vented Euphonia it was fifteen days, and in the same tree in a different year it was seventeen days. Eggs of the Yellow-throated Euphonia hatched sixteen days after the last was laid. At one nest of the Tawny-bellied Euphonia the interval was between sixteen and a half and eighteen days, and at another it was eighteen days. These are the longest incubation periods determined for any tanager; the substantially larger eggs of the Scarlet-rumped Tanager and its congeners hatch in twelve days plus a few hours. Since euphonias incubate with normal constancy, one wonders why their eggs take so long to hatch. In other families, notably the American flycatchers, some of the smallest species have much longer incubation periods than larger species. Could the explanation be that smaller eggs cool more quickly during the absences of the female, who incubates with no help from a mate, and this retards the development of the embryo? Another suggestion is that the length of the incubation period is correlated with the length of the nestling period. Although nidicolous nestlings often flutter from low, open nests when they can hardly fly, it is advantageous for the young to remain in high, covered, relatively safe nests until they can fly well. Perhaps the euphonias' incubation period is long because their nestling period is long.

The newly hatched nestlings are blind, with such short, sparse down that, unless carefully examined, they appear quite naked. The skin of Tawny-bellied Euphonias is pink, but that of White-vented Euphonias is blackish and shiny. In these and other euphonias, the interior of the mouth is red, as is usual in tanagers. The flanges at the corners of the mouth may be white, as in the White-vented Euphonia, or yellowish, as in the Tawny-bellied Euphonia. The parents promptly remove the empty shells.

Although other tanagers bring food visibly in their bills, when euphonias come to feed their nestlings they appear to have nothing for them. Nearly always the two parents arrive together, and the male goes first to deliver his food. Resting in the doorway, head inward, he regurgitates to the nestlings, then waits nearby while their mother feeds them. If she does not stay to brood, the two fly away together. The male's almost invariable habit of feeding first, which I have seen at nests of four species of euphonias and is also true of chlorophonias, saves unnecessary movement while the nestlings are only a few days old and need covering. If the female went first to the nest, she would have to withdraw so that her partner could occupy the doorway, then return to brood. This order of feeding, begun as soon as the nestlings hatch, persists after they no longer require brooding, when it would appear to make no difference which parent fed first. Sometimes the male, after feeding the nestlings and returning to a perch, closely escorts his mate as she goes to feed, just as he had done when she entered to incubate. She may also fly beside him as he goes to the nest, so strong is the euphonias' habit of accompanying their partners.

At a nest of White-vented Euphonias that I could watch through the doorway of my study, three mouths opened like red flowers whenever a parent alighted in front of them. Promptly the adult proceeded to regurgitate food and place it in these corollas, a little at a time, first in one and then in another. The parent's short bill was not inserted in a nestling's throat, as in hummingbirds. To regurgitate, the male retracted his neck and jerked his head up and down; the female twitched her head sideward. Looking sharply, I could see a bulge traveling up her foreneck. On a single visit to the nest, each parent delivered the food in from four to fifteen portions, mostly from eight to twelve. Sometimes a nestling received several successive portions, but usually all three were fed on each parental visit.

As in pigeons, the time that a parent took to deliver a meal to the three nestlings decreased as they grew older, from fifty to ninety-four seconds at five days of age to thirteen to thirty-five seconds at fifteen days. The number of times each parent visited the nest to feed increased only slightly, from twenty-one in six hours for three nestlings four and five days old to twenty-eight in six hours when they were fifteen and sixteen days of age. This was only little more than one and a half feedings for each nestling in an hour, which is a low rate for passerine birds; but when nestlings are fed by regurgitation, their apparently more copious meals tend to be more widely spaced than when they are fed directly from the bill. After a meal, a large, dark swelling on the front and right side of a nestling's neck, as in hummingbirds and others fed by regurgitation, revealed where the food was stored awaiting digestion.

A male Tawny-bellied Euphonia appeared to have trouble finding his newly hatched nestlings. He would accompany their mother to the nest, and while she fed them he would flit around the nest as though searching for them or hover momentarily in front of it, then fly away without having given them anything. By their third day, however, he was feeding as male euphonias usually do. Sometimes he brought the day's last meal when the light was already so dim that I could hardly distinguish him as he clung in front of the nest. The female of this pair sometimes delivered as many as thirty portions at a single meal and her mate up to eighteen. When their two nestlings were two days old, their mother brought nine meals in six hours; their father, none. At nine and also at sixteen days, each parent brought food eight times, making sixteen meals in six hours, at the rate of one and a third meals per nestling per hour.

After delivering a meal, the parent euphonias attended to sanitation. Although the droppings of most passerine birds are enclosed in white, gelatinous sacs that facilitate their removal, parent euphonias enjoy no such convenience. At the termination of a feeding, the parent would wait at the doorway until one or more nestlings turned around and presented their rear ends, from which oozed a gelatinous or watery stuff containing small seeds and fragments of tiny insects, which the parent ate little by little, appearing almost to lick it up with its tongue. Sometimes the male, after waiting nearby while his mate fed, would return to the doorway beside her and help remove the excrement. I never saw either of them carry away anything in their bills. Sometimes the fecal matter pulled out in long, clear strings, like the droppings of adult euphonias. This colorless, viscid matter was probably from seeds of mistletoes, which the parents gathered after they flew beyond view.

The nestlings were brooded very little. Two-day-old Tawny-bellied Euphonias were covered by their mother for little more than one-third of the first six hours of the day. A female white-vented Euphonia brooded nestlings four and five days old for a total of only one hour during six hours of a sunny morning. Two broods of the latter and one of the Tawny-bellied Euphonia were covered at night by their mothers only until they were eight days old. Although the roofs of their nests gave some protection from the frequent rains of this season, to leave nearly naked nestlings exposed to the often chilly nocturnal air at 2,400 feet (730 meters) above sea level seemed rather harsh treatment. However, when alone in their nest after nightfall, eight-day-old nestlings felt warm, and they appeared not to suffer from exposure. Such early cessation of nocturnal brooding is widespread among birds with covered nests, and hummingbirds are left exposed in open nests when little older. Most passerine birds with open nests continue nocturnal brooding until their nestlings are well feathered.

Nestling euphonias develop slowly. At eight or nine days, their eyes begin to open, and by the following day they are fully open. By their ninth day, the pins of their wing plumes have become long but their body feathers are just pushing through the skin. At twelve days, when larger tanagers such as the Scarlet-rumped leave the nest well clothed, euphonias are still nearly naked. At thirteen days, their plumage begins to escape its horny sheaths, and when fifteen days old they are fairly well covered, except on the head. A brood of three Yellow-throated Euphonias left the nest when seventeen days old. Two Tawny-bellied Euphonias departed at the age of twenty days. Three White-vented Euphonias that hatched in seventeen days also remained in the nest for twenty days; but three of a different brood that hatched in only fifteen days left when eighteen and nineteen days old. Perhaps the unusually dry weather of the June when this brood was raised, in the same tree where the preceding nest was located, was responsible for its more rapid development. For Yellow-crowned Euphonias, I have recorded nestling periods of twenty-two and twenty-four days. All these nestlings were dull olive-green above and yellowish below, somewhat like their mothers; but the young Tawny-bellied Euphonias lacked the chestnut forehead and tawny abdomen of the adult female. I have not known either a parent or newly emerged young to return to sleep in the nest, as certain other birds with roofed nests do.

The day after the departure of three young White-vented Euphonias, their father repeatedly looked into the empty nest. On the following day, he remained inside for a minute or two, something I had never seen him do while the nest held eggs or nestlings. A few days later, he was entering at intervals in both the morning and the afternoon, singing his artless song while within. After this had continued for a week, he carried in a tuft of moss, and on the next day he brought more materials. Now the female, who had probably been too busy with her fledglings to give much attention to the empty nest, entered to shape it. They continued to refurbish the old nest; and sixteen days after the departure of the first brood of three the female laid the first egg of another set of three, which vanished before they hatched, possibly taken by a squirrel. Seventeen days after a brood of two Tawny-bellied Euphonias flew from their nest, the first egg of another brood was laid in it. Only eight days after a single young Yellow-crowned Euphonia departed its nest, the structure had been completely relined and contained a newly laid egg of another brood. Although euphonias seem frequently to try to raise second broods, I have not known them to nest successfully twice in a year. A female Tawny-bellied Euphonia who lost her eggs from a nest hanging above a river used it again in the following year, but in both years she lost her eggs.

The White-vented Euphonias that I watched through the doorway of my study nested in the Calabash tree where the Tawny-bellied Euphonias had long been sleeping. One of the two male Tawny-bellied Euphonias who slept on the same long branch where the White-vented Euphonias were building tried to drive them away; but they finished their nest, laid three eggs, and successfully raised their brood. Late on many an afternoon while the pair nested, the Calabash tree was the scene of a lively, largely vocal confrontation between the males of the two species. For many minutes and sometimes for as long as three-quarters of an hour, the two yellow-breasted birds called against one another. The male White-vented Euphonia repeated a high-pitched note that resembled the Tropical Kingbird's call, a sharp *chip*, another high note, and a lower, almost whistled note. The male Tawny-belly called *r-r-r-ra cha cha* or *cha cha, cha cha*. While they vociferated, both turned from side to side, flipped their wings with a slight, rapid movement, and twitched their tails. At intervals the Tawny-belly scratched and preened. Sometimes he chased the smaller White-vented Euphonia, but I did not see them clash together.

One evening, after he had retired early into his sleeping niche, the Tawny-bellied Euphonia emerged and five times drove the female White-vented Euphonia from near the nest that she was trying to enter. After some of these attacks, he looked into the nest without harming the nestlings. On several occasions he chased away the male White-vented Euphonia while the latter was feeding his nestlings. After one of these altercations, the two parents fed while standing side by side in the doorway, an unusual occurrence caused by their excited state. Sometimes the Paltry Tyranniscus who was feeding nestlings in this tree and the Bananaquit who slept in it were interested but passive onlookers of the dispute, and once a Little Hermit Hummingbird hovered in front of the vociferating Tawny-belly. Finally, he would fall silent and slip into the pocket amid the liverworts where he slept. Soon the White-vented Euphonia stopped chattering, too, and in the failing light flew high in the air toward the forest to roost where I could not find him. Then all became quiet in the tree where the female White-vented Euphonia brooded her nestlings and the tyranniscus warmed hers in a cozy covered nest tucked among the liverworts.

Another cause of excitement was the occasional visit of a second pair of White-vented Euphonias. When they approached the nest, the resident male flew at the intruding male, the resident female at the visiting female. These pursuits were mild, the birds did not strike each other, and after a little chasing the four would perch for a while not far apart in the nest tree. The visitors soon flew away. Not a feather was lost in any of these tiffs among euphonias.

Euphonias are not as numerous as they used to be, and a decade has passed since they last nested near our house. In addition to destruction of habitat, their rarity is caused by illegal trapping, for male *monjitas* (little nuns), as they are called in Costa Rica, are favorite cage birds here as in other countries of tropical America. The unhappy birds are usually confined in cages so small that they cannot fly, they are fed inadequately, and they rarely survive long. It is distressing to think how heavily birds are penalized for being lovely, songful, and harmless.

27. *The Emotions of Parent Birds*

Often, while I watched birds expertly building their nests, patiently incubating their eggs, busily feeding their nestlings, or protecting them from hot sunshine, cold rain, or menacing predators, I have wondered what emotions accompany these so varied activities. Do they feel anything like satisfaction or pride in a well-made nest, any stirrings of anticipation as they warm their eggs to life, any affection for the young they nourish and defend, any anger when some predatory animal or intruding human threatens their brood? Or is the whole complex sequence of activities whereby they reproduce their kind simply the unconscious performance of operations programmed in their genes, so that, as some have maintained, they may be regarded as living mechanisms, extremely sensitive to stimuli but devoid of foresight and sentiments? Do they sing without joy and vigorously protest threats to their young with no trace of anxiety?

These questions, which have exercised the minds of many sensitive observers of animal life, are certainly not easy to answer. Intuitively, we ascribe to creatures that resemble us in so many ways, that belong to the same vertebrate stock that produced us, emotions like those that we would feel in similar circumstances. Especially in childhood, before thought has become critical, we firmly believe that the animals around us have minds rather like our own. But how shall we convince the skeptic, who can argue plausibly that the whole Universe and all it contains is no more than an idea in his own mind, without real external existence? How shall we prove to him, who can so deftly brush aside the persistent testimony of his five senses, that feelings somewhat like his own stir in the hidden depths of beings whose very existence he professes to doubt?

Beyond the extremes of skepticism, which skeptics themselves do not permit to interfere with the serious business of daily life, is the sobering thought that one major aspect of reality, the psychic, which may well be coextensive with the mate-

White-necked Jacobin Hummingbird

rial aspect, is concealed from us, except as manifested in our individual selves. Lacking a sense like sight or hearing that reveals it directly to us, we can only infer its presence from indications that these senses do bring to us, the vocal or visual expressions of the emotions such as Darwin studied in perceptive detail. Despite the ease with which we can be misled and the risks of misinterpretation, it is hardly less than our duty to make every effort to penetrate the veil that hides from us the psychic life of the creatures that surround us. Until we gain better understanding of this aspect of reality, not only shall we lack guidance in our dealings with the animals around us, we shall lack profound insight into the nature of the Universe and our place in it.

The major activities of nesting birds, indispensable for rearing their broods, afford no ground for inferring emotion. Their uniformity in a species over wide areas leaves no doubt that they are innate, programmed in their genes. Conceivably, they could be performed by a mechanism devoid of feeling, but one much more adaptable to the opportunities and obstacles of the environment, with greater powers of adjustment to unpredictable circumstances, than machines that man has invented. Probably a bird who builds an elaborate nest with a variety of materials, as many tropical and some temperate-zone birds do, is guided by something like an innate image of the finished product, so that the several stages in construction follow in due order and the builder knows when to stop adding piece to piece; but we cannot prove this. The demonstration that the songs of birds serve the quite utilitarian purposes of proclaiming possession of a territory and attracting a mate forbids us to regard them as simply emotional outpourings or expressions of high spirits. Nevertheless, it is difficult to believe that birds who invent scores of songs, as some wrens do, or display great skill in mimicry, like mockingbirds and others, do not delight in their virtuosity. The avoidance of monotony by even less gifted songsters seems to reveal a modicum of aesthetic feeling, as Hartshorne (1973), the philosopher-ornithologist, maintained. The elaborate mutual displays of birds like herons and grebes, which to Huxley (1923) were revelations of strong emotions, serve to synchronize their sexual responses and insure that the female's eggs will be fertile; can we be sure that their significance is more than physiological?

Perhaps, in our groping efforts to explore the emotions of birds, we should ignore the essential and turn our attention to the unessential. The activities indispensable for the preservation of the individual and the species have been perfected and reinforced by natural selection, a rigorous process that, apparently, has little regard for feeling. But, while he watches birds engage in these scheduled activities, the attentive observer sometimes hears or sees them do things that, for all he can tell, are wholly superfluous. Sometimes, indeed, they seem not only unnecessary but rash, because they might draw a predator's attention to their eggs or young. Accordingly, far from promoting them, natural selection should suppress them or at least hold them within limits that do not jeopardize progeny. These little sounds or acts, these overtones or embellishments of the serious business of living and reproducing, appear to be spontaneous expressions of feeling, of no "survival value," except insofar as strong emotion may increase the intensity of effort. In birds, as in ourselves, voice serves both to communicate and to express emotion. When no attempt to communicate is evident, the sounds that we hear are probably expressions of feeling. In this chapter, I tell about certain activities of birds, especially their vocalizations, that appear to be, above all, revelations of emotion.

I shall omit the courtship ceremonies

that stir some of the larger birds to the highest pitch of excitement. With few exceptions, the small birds of tropical forests and fields amid which I have lived for many years form pairs in a manner so inconspicuous that it has escaped me. (The more spectacular displays of male manakins, hummingbirds, and others that form no lasting bonds with the other sex need not detain us here.) The many tropical birds that live in pairs throughout the year, probably with lifelong fidelity, maintain close contact with their partners by calling or singing to each other (antiphonally in the case of certain wrens), by mutual preening, or by gifts of food. Usually the male feeds his mate, but occasionally she returns the courtesy, as in euphonias. Exceptionally, the constantly mated birds practice little greeting ceremonies. Whenever the Black-striped Sparrows that hop over our lawn and share the chickens' corn come together after a short separation, they greet each other with a brief, rapid flow of unmelodious notes that I hear only on such occasions. As the members of a pair of Vermilion-crowned Flycatchers approach each other, both flutter their wings while they deliver a short, high-pitched, somewhat trilled salutation, which they may repeat at intervals while they perch not far apart in a treetop. This soft, chiming utterance serves both as a greeting and as an assurance that each is within hearing of its mate and all is well with them. Whatever emotions stir the breasts of these constant companions appear to be enduring rather than intense.

Some of the American flycatchers are especially emotional; at least, they express their feelings more freely than do many of the birds amid which they dwell. While seeking sites for their nests, they sit in promising spots in trees or shrubs while they voice a continuous flow of soft, twittering notes that I have called a "nest song." Among the flycatchers from which I have heard this utterance are the Vermilion-crowned, the related Gray-capped, the Sulphur-bellied, Golden-bellied, and Boat-billed. Sometimes this nest song stimulates neighboring pairs to rest in potential sites and sing. In the trees around our house, I have heard three species voicing these pleasant twitters at the same time. Since the sexes of flycatchers are usually alike in plumage, it is not always possible to tell which sings, but I am certain that sometimes a male sits and twitters in a newly started nest that he will not help build. The emotions of these birds about to engage in what is probably a satisfying activity are contagious, pleasantly exciting not only to their feathered neighbors but also to us who are interested in them.

While we sat at breakfast before sunrise on the last two mornings of April and the first of May, a Garden Thrush flew repeatedly to a certain spot amid the large red-and-green leaves of a *Taetsia* shrub just outside the dining room window. Here he stood for a minute or two while he sang a few subdued notes. After he had sung here on three mornings, he and his mate started a nest in the spot that he had chosen. Although it is not usual for male Garden Thrushes to help build the substantial, mud-walled cup, he continued to bring many billfuls of material to it.

In an earlier April, a male Scarlet-rumped Tanager came again and again to sing in the bushy *Thunbergia* in front of my study window. After he had done this for several days, his mate went repeatedly to sit in the same spot amid the close-set twigs while he perched, often singing, below her in the shrub. On the following day, she started a nest in the site that he had selected with song and, without his aid, finished it in four days of leisurely building. Although female Buff-throated Saltators are usually songless, once I heard one sing in a whisper while she gathered material for her nest.

Incubation often impresses us as

stolid, emotionless sitting which the more active of us would find excessively boring. However this may be with a broody hen who, despite our efforts to discourage her, continues for days to sit upon an empty nest, some birds evince emotional attachment to their eggs. Again, it is the active flycatchers who most clearly reveal that incubation is something more than the transfer of heat from the featherless incubation patch on the bird's abdomen to the developing embryo beneath her. While sitting in a blind before a flycatcher's nest or just passing by, I have frequently heard the subdued nest song, which reminded me of a human humming to herself while contentedly engaged in her daily tasks.

After most of her absences to seek food, a Vermilion-crowned Flycatcher sang a longer or shorter nest song as she passed through the round doorway in the side of her covered nest and snuggled down on her two speckled white eggs. Sometimes she twittered softly in the midst of a session of incubation; at intervals she called more loudly in varied tones and was answered by her mate. While she called in the nest, her usually concealed crown patch, spreading like a brilliant scarlet flower, seemed to illuminate the whole interior of her ample chamber. Flycatchers of many kinds nearly always keep their red, yellow, or white crown patches covered except when stirred by emotion. A neighboring Vermilion-crown was far more silent, entering her nest with a subdued version of the nest song but rarely repeating it while she covered her eggs. During her absences, her mate, who never incubated, sometimes flew to the doorway and, resting there with his head inside, sang with low, soft notes rapidly repeated. He, too, appeared to be emotionally stirred by the sight of the eggs.

Little gray Lesser Elaenias often appear to find incubation an irksome task, for during their usually short sessions on the nest they constantly turn their heads from side to side and look around. Nevertheless, while sitting they may continue for minutes together a low soliloquy of short notes joined in a quaint little nest song suggestive of perfect contentment. Other flycatchers that twitter softly while incubating include the Gray-capped, Dusky-capped, and Boat-billed. Even the pesky Piratic Flycatcher, who calls so annoyingly while trying to win a nest built by a more industrious bird, voices a low, melodious, rhythmic murmur that may become a soft warble while she sits calmly on her eggs in a stolen structure, which is always closed or roofed. The vociferous Bright-rumped Attila has sometimes been included in the cotinga family and sometimes with the flycatchers. Like the latter, the female voices a subdued nest song while she incubates her three or four mottled eggs in a niche in a trunk or amid epiphytes.

I have not heard the twittering nest song from true songbirds, among which males and even females may sing more loudly while they incubate. For nearly two hours, I watched a male Rufous-browed Peppershrike sit in a high, inaccessible nest without ceasing to repeat, at intervals of a few seconds, his loud, far-carrying song. Among the related vireos, males, including the Yellow-throated and the Warbling, likewise sing while they incubate. The full-throated songs of incubating male Rose-breasted and Black-headed grosbeaks guide ornithologists to their nests. Among females that sing loudly while they incubate, in response to their mates or otherwise, are Highland Wood-Wrens, Melodious Blackbirds, Yellow-tailed Orioles, and Blue-black Grosbeaks. In all these birds, the need to express emotion appears so strong that it overrules prudence.

Likewise, when approaching their nests, some birds are needlessly noisy, moving around while they repeat loud notes, expressing excitement or anxiety at the risk of attracting the attention of the predators that they appear to fear. Even in predator-ridden tropical forests, where birds should return to their nests with silent directness, species such as the Black-hooded Antshrike, Blue-black Grosbeak, and Red-crowned Ant-Tanager appear unable to suppress their emotions.

Some male birds who do not incubate reveal their continuing attachment to the nest by bringing food to it more or less frequently before the eggs hatch. Among jays, goldfinches, siskins, crossbills, and hornbills, the food is obviously intended for the female and sufficiently abundant to increase substantially the time she can remain incubating without going to forage. Often, however, the food is brought in the female's absence and presented to eggs that may be still far from hatching. With laden bill, the impatient father lowers his head into the nest and utters soft notes, as though coaxing nestlings to lift up their open mouths and eat. I have seen such anticipatory food bringing in birds as diverse as tanagers, wood-warblers, flycatchers, and tityras. If the female happens to be present when her partner comes with food for nestlings still unhatched, she may receive or ignore it, as though disdaining to accept what was not intended for her. More rarely than males, I have seen females bring food to their eggs. The feeding of nestlings may be no more than an innate response to their highly colored gaping mouths and begging cries, but this can hardly be true of anticipatory food bringing, which is too sporadic and unnecessary to be genetically programmed but appears to reveal a strong affective attachment to the nest and alert interest in what is happening there.

Among the birds that each year nest in our garden is a tiny gray birdling to which an ornithologist who did not appreciate it gave the uncomplimentary name *Tyranniscus vilissimus*, which has been translated as Paltry Tyranniscus. Amid the profuse growth of moss or liverworts hanging beneath a slender branch of a Calabash tree, the female builds a little mossy globe with a side entrance. One April afternoon, while I set up a stepladder to look into one of these nests, the female tyranniscus clung at the doorway, head inside, repeating notes so soft and low that I would not have heard them if she had not been so close above me. After continuing this subdued song for possibly thirty seconds, she climbed inside and settled down to brood, with her head outward. Evidently, she had been so absorbed by what was happening in the nest that she failed to notice me set up the ladder, for nearly always the tyranniscus is more wary. Presently, becoming aware of her visitor, she flew out and complained with the low, mournful notes usual on such occasions. Looking into the nest, I saw a nestling in the act of hatching, a little of its tiny pink body visible between the separating parts of the shell. Its mother had been so strongly moved by the sight that she forgot her customary caution, as birds rarely do.

A female Lesser Elaenia responded differently to the birth of her brood. As she sat on her nest in a Rose-apple tree beside the house in the dim early light of an April morning, with a newly hatched nestling and a hatching egg beneath her, she continued for many minutes to deliver a subdued, simplified version of the male's dawn song. *De-weet, de-weet* she repeated countless times in a queer, dry voice, surprising me greatly, for I believed that this utterance was confined to males, and during the fortnight while she incubated so close to an open window that all her notes were audible to me while I sat at the dining table, I had heard nothing like this from her—only her call notes and her very different nest song. The advent of her nestlings had stirred her to voice notes that she seldom uttered. While a Boat-billed Flycatcher incubated in a guava tree close beside the house, she was not heard to voice the nest song that she had frequently used while selecting the site of the nest and building it; but on the day a nestling hatched she was moved to utter these low, soft notes again. It might be thought that these vocalizations just as the nestlings hatch serve to advise their father of the event so that he will promptly begin to feed them, but a careful study of this point yielded no evidence for this (Skutch 1953, 1976).

While their eggs hatch, birds sometimes behave unpredictably. Like other antbirds, male and female Black-hooded Antshrikes alternate on the nest by day and the female occupies it by night. Nevertheless, with an egg hatching beneath him, a male antshrike, who had already been sitting for an hour and a half, refused to relinquish the nest to his mate. Later, at her first opportunity to see the nestling, she stood for several minutes on the nest's rim, silently contemplating it, as parent birds frequently do when they first see a hatchling.

While nestlings hatch beneath them, some females sit with normal constancy, but one Orange-billed Nightingale-Thrush remained continuously on her hatching eggs for four and a half hours, more than twice as long as any other session that I have timed in sixty hours of watching at eight nests of this species. She sat nearly motionless, neither fidgeting nor rising to look beneath herself, as some birds do while their eggs hatch. Quite different was the behavior of another Orange-billed Nightingale-Thrush, who, instead of covering an emerging chick, stood on the rim of her mossy nest with her head lowered into the bowl, intently watching and perhaps assisting the hatchling's escape from the shell. Both parents started to bring food before the chick was able to eat. With patient coaxing to the accompaniment of soft, encouraging notes, they tried to induce the newly hatched nightingale-thrush to take its meals; but together the parents brought more than it and its mother could consume, with the result that some of the tiny insects were carried away uneaten.

Male birds who do not incubate sometimes respond emotionally to the first sight of their nestlings. With subdued notes that resembled the nest song, males of both the Gray-capped and Vermilion-crowned flycatchers brought their first offerings to hatchlings, which they did not discover until a few hours after their mother had begun to feed them. In these and other flycatchers, the delivery of food to nestlings during their first few days is often accompanied by a soft feeding song, not greatly different from the nest song that is heard while the adults build and incubate.

When we approach a nest that contains eggs or young or perhaps from concealment watch while a predator comes within view, we frequently witness behavior that could hardly have been favored by natural selection because, far from promoting the nest's safety, it may jeopardize the nest by calling the predator's attention to the fact that the birds have eggs or young nearby. Parent birds who disappear in discreet silence at the approach of an enemy too formidable to confront behave more wisely. The cries plaintive or harsh, the futile flitting around, of parents whose nests are or appear to be in peril can hardly be other than expressions of distress without "survival value." Garden Thrushes are especially prone to complain at the least hint of danger to their young. I have walked through plantations and bushy pastures where many of these plain brown birds nested to the accompaniment of the querulous cries of a sequence of nervous parents that finally grated on my nerves.

Hurried snatches of song reveal the nervous tension of other parents concerned for the safety of their young. A Black-cowled Oriole and a White-collared Seedeater sang while I inspected their nests, and one of the latter burst into fullest song while he tried to repulse a predatory Kiskadee Flycatcher. A Variable Seedeater and his mate mingled song with mournful cries while a slender snake swallowed one of their nestlings.

The height of imprudence is shown by birds who try by direct attack to defend their nests from animals much more powerful than themselves, like the male Slaty Antshrike and the female Gray Catbird who boldly pecked the hand that I placed upon their nests, or the female Scarlet-rumped Tanager who was caught by the snake that she tried to repel from her eggs and hung helplessly by one wing until I released her. Such rash displays of strong parental feeling should be repressed by natural selection, for the death of the parent would nearly always be followed by that of the progeny. Prudent parents who, by timely withdrawal, save their own lives so that they may try again, perhaps with better luck, to raise a brood are more likely to transmit their genes to future generations.

The distraction displays of birds driven from their eggs or young by an approaching animal are just the sort of behavior that we would expect of a bird forced to abandon a nest to which it has a strong emotional attachment. Typically, the incubating or brooding parent clings to its nest to the last moment compatible with its escape. Then it flutters or drags itself over the ground, wings spread and flapping loosely, as though crippled and unable to fly. If one follows, it barely manages to keep ahead of its pursuer until it has led him a good distance from the nest, when it suddenly "recovers" and flies lightly away. The simulation of helplessness is often so realistic that it has led naturalists to conclude that, torn between parental attachment and the instinct to survive, the bird is suffering a fit and is temporarily incapable of coordinated movement.

However, we have good reasons to believe that, far from being convulsed, the injury-feigning bird is in full possession of all its faculties. In the first place, it rarely gives this act unless it has an adequate stage, such as open ground where it is in no danger of becoming entangled in dense vegetation and caught by its pursuer. If a patch of tangled vines lies in the path of the fluttering bird, it may fly over it and resume its display in a clear space on the farther side, like the White-tipped Dove in chapter 2. And it manages to keep just far enough ahead of the pursuing animal to lure it onward with the prospect of an easily captured meal without incurring the risk of being seized. Clearly, the seemingly crippled bird has command over all its movements and is fully aware of the risk it takes.

Although at present injury simulation is one of the innate behavior patterns of parent birds, like nest building and feeding the young, perhaps it originated from emotional conflicts and convulsive behavior of birds driven from nests to which they were strongly attached. If reluctant, limping withdrawal from endangered nest or young often lured the enemy away and saved the progeny without loss of the parent, natural selection would favor such behavior until by slow degrees it became innate and, no longer dependent upon disabling emotional conflicts, more stereotyped and effective. So widespread are these displays today, especially among birds that nest on or near the ground and those that lead precocial chicks over the ground, that this process of replacing paroxysmal behavior by genetically programmed behavior, an example of organic selection or genetic simulation, must have occurred among ancestral birds very long ago and probably repeatedly in different lineages.

Although some birds appear to take their fledglings' or chicks' departure from the nest as a matter of course, just another step in the strenuous routine of rearing a family, others are stirred emotionally. Years ago, I passed four mornings watching for the departure of a young Vermilion-crowned Flycatcher who remained exceptionally long in its roofed nest in a lemon tree. Soon after sunrise of the fourth morning, while its parents were absent, it flew from the nest to the leafy crown of a Madera Negra tree down the slope. When a parent—the mother, I believe—returned with laden bill and found the nest empty, she flew with her mouthful of insects to the crown of a guava tree at the foot of the steep slope, thence back to the nest tree, from which she went to another guava tree near the first. Here she delivered a long sequence of soft, low notes hardly to be distinguished from the nest song. Promptly, the fledgling flew from the Madera Negra tree to the guava tree where its parent sang with her bill full of food, which doubtless it received amid the concealing foliage. She had called it from the exposed position where it first alighted to one well screened on all sides. Then for many minutes the parent Vermilion-crown continued to pour forth her low, sweet notes amid the sun-drenched foliage of the guava tree with her fledgling beside her. Its departure, marking the successful conclusion of the nesting, stimulated a much longer flow of the same song that had accompanied the selection of the nest site and the start of building nearly two months earlier. Similarly, the departure of a fledgling Gray-capped Flycatcher prompted many repetitions of the parents' nest song, which I had not heard during their busy days preceding its emergence from the nest.

While I stood in a seldom-used Venezuelan roadway watching a pair of Pale-headed Jacamars catch skipper butterflies and dragonflies for their brood of four in a burrow in the vertical bank below me, one of the fledglings emerged. Already well feathered, it easily flew up to the crown of a tree in front of me, where both parents immediately joined it and started to sing. Each crescendo of sharp *weet*'s and twitters led up to a high, thin trill. With variations, they repeated this song again and again while their bodies turned from side to side and their tails twitched rapidly up and down, as though beating time to their animated notes. At intervals the fledgling joined in with its weaker voice, flagging its tail, as its parents did. For a long while, the jacamars continued to sing what seemed to me a hymn of triumph, celebrating the successful conclusion of their nesting or perhaps congratulating the fledgling on its graduation from the burrow. The exit of another fledgling later that same morning was greeted with song by the single parent who promptly alighted beside it high on a spray of bamboo, but this time the performance was briefer, as though the prolonged outburst that celebrated the emergence of the first fledgling had drained the jacamar of emotion.

The emergence of two young Rufous-tailed Jacamars from a burrow in a low Costa Rican bank was accompanied by much song by their parents as well as the fledglings themselves, who while still in the nest had started to sing weaker versions of the adults' elaborate songs. Southern House-Wrens, unquenchable songsters, sing more profusely just before and during their fledglings' departure from a gourd, nest box, or cranny in a building or tree. Birds so diverse as Black-crowned Tityras who raised three fledglings in a hole high in a trunk and Marbled Wood-Quails who hatched four chicks in a covered nest on the ground seemed unusually excited as their young were about to fly or walk from the nest.

Nests where a brood was raised may continue to stir the parents' emotions. Throughout the week after the emergence of their two fledglings, both parent Gray-capped Flycatchers, singly or together, went repeatedly to their empty nest to deliver the nest song or sundry spirited or bizarre sequences of notes, for these flycatchers have a varied vocabulary. Sometimes they brought food to the empty nest before taking it to their fledglings outside. Even a full month after the young birds flew, I watched both parents "sing" loudly near the nest. During all this interval, they dutifully attended their

Pale-headed Jacamar

young in neighboring trees. I surmised that this sustained interest in the nest would lead to a second brood, but in this I was mistaken—second broods are rare among Gray-capped Flycatchers in this region.

In chapter 26 I told of a male White-vented Euphonia's strong interest in the nest that his fledglings had just left. Likewise, male Scarlet-rumped Tanagers sometimes sing beside nests where their young were raised. For over a month after the fledglings departed from a nest amid the variegated foliage of a croton shrub beside my study, a male, apparently their father, came almost daily, sometimes twice a day, to sing profusely while standing in the empty bowl, on its rim, or close beside it. Finally, forty-six days after the first brood flew, his mate started another nest a few yards from the old, disintegrating structure.

These and other incidents that I have witnessed in more than half a century have left a strong conviction that a bird's nest and its contents are more than "releasers" that set off integrated sequences of genetically determined activities but are objects of strong emotional attachment, attended with fervor and, when feasible, protected with zeal. Possibly because of the perils that surround them, the constantly paired birds of tropical woods and gardens that I have watched may never know the highly charged emotional states suggested by the mating ceremonies of certain larger birds of more open places. Their more subdued emotions, stirred by their nests and young, are most clearly revealed at the critical points of a nesting sequence: selection of a site and start of building, hatching of the eggs, and departure of the young.

Birds impress us as being more emotional than intelligent. It is widely believed that intelligence is rather closely correlated with the size of the brain, but recent experience with computers, which can be made more capable as they are made smaller, casts doubt upon this. It appears that our own brains might be substantially reduced in size with no loss of intelligence. The quality of a brain seems to be much more important than its size. The evolution of intelligence is closely linked with an animal's ability to perform useful operations with its limbs, whatever form they may have. Long ago, Aristotle recognized that man is the most intelligent animal because he has hands. His mind evolved rapidly as the hands that it directed became increasingly efficient in activities that promoted survival. Similarly, a bird's intelligence appears to be nicely adjusted to its capacity to do things with bill, wings, and feet. It can forage efficiently, build charming nests, take excellent care of its young, and navigate between pinpoints on the map thousands of miles apart in a manner that we fail to understand completely; but it cannot do nearly as much as we can do with a pair of five-fingered hands. If birds had more intelligence, they might not be able to use it for lack of means to execute their ideas.

Indeed, to have intelligence far beyond the ability to act might induce a debilitating sense of frustration, such as we sometimes feel when deficiency of strength or skill prevents the accomplishment of what we know we should do. Birds, fortunately, are spared the distress of having foresight and intelligence disproportionate to their practical ability. They appear not to anticipate disasters that they lack means to circumvent or to make plans that they cannot execute. Like little children, they appear to feel more than they think, and they are blessed with the enviable faculty of soon forgetting the disasters that so frequently befall their faithfully attended nests and starting afresh to raise a brood that will engage their affections.

Epilogue

I began this book with reflections on light and space; the last chapter considered an aspect of consciousness. The choice of subjects was not accidental, for space and consciousness appear to be the alpha and omega of existence. Space is the container and matrix of all that exists, the continuum that binds all things together and makes this a Universe rather than a loose collection of unrelated things. Consciousness appears to be the end and goal of the Universe, which it strives to increase and intensify, because without feeling and awareness it would exist barrenly, devoid of value. Although space and consciousness seem to stand at opposite poles they are probably much more closely linked than we realize.

Nothing is more familiar to us than space and consciousness. In space we live and act and keep our possessions. Consciousness declares its presence in us all our waking hours and even while we dream. The skeptic who questions the reality of everything external to his mind cannot doubt that he is conscious. Nevertheless, paradoxical as it may seem, these two familiar things are the least understood, the greatest of the mysteries that surround us. In the first chapter, I did not attempt to explain how space conveys light waves over vast distances or how it draws all things together. Although now we measure with great accuracy the strength of the gravitational field and the velocity of falling bodies, how gravitation works—its mechanism—is as great a mystery to us as it was to the ancients who first tried to explain why some things fall and others rise. In the last chapter, I could not prove that birds are emotionally attached to their nests and feel strongly about their young; I could only present my reasons for believing that they do.

Mind—consciousness—is revealed directly to us only in our individual selves; its presence anywhere else, even in the people closest to us, is only an inference or an intuition. The distribution and intensity of feeling in our planet, or even in the living world, have never been mapped as we map its magnetic fields. Even in ourselves we do not know how it is related to the body; introspection hardly reveals where it is. The ancients, including one of the greatest of all thinkers, Aristotle himself, regarded the chest rather than the head as the seat of feeling and thought. If now we are certain that we think with our heads, it is probably because anatomy points to this conclusion and we are taught from childhood that this is true, rather than a spontaneous revelation.

Compared with the vast bulk of our information about matter, its composition and behavior in the most varied conditions, what we can say about space is meager indeed. Although psychologists tell us much about the human psyche, they have not probed it to its prime foundations, and, after all, man's mind appears to be only a small province in the whole realm of consciousness. Ethologists describe in great detail how animals behave but they do not reveal what they feel and think. To understand this perplexing Universe, we need to know far more than we do about space and consciousness, its alpha and omega.

These limitations of our understanding should challenge rather than oppress us, make us humble but not despondent. Between these two enigmas lies an infinity of things to delight and exercise our minds, in all parts of Earth where life abounds but nowhere more than amid tropical profusion. As, year after year and generation after generation, we sharpen our minds and deepen our insights, we should come closer to understanding the ultimate mysteries. This, at least, is the faith of one who, in a contemporary world that contains much to oppress us, has been sustained by the fascination and splendor of tropical nature.

Agouti

Bibliography

Baker, H. G. 1970. Two cases of bat pollination in Central America. *Rev. Biol. Trop.* 17: 187–197.
Beehler, B. 1980. A comparison of avian foraging at flowering trees in Panama and New Guinea. *Wilson Bull.* 92: 513–519.
Bent, A. C. 1948. Life histories of North American nuthatches, wrens, thrashers and their allies. *U.S. Natl. Mus. Bull.* 195: i–xi, 1–475.
Chakravarthy, A. 1954. Jain ideas in the modern world. *Aryan Path* (Bombay) 25: 461–464.
Chapman, F. M. 1929. *My tropical air castle: Nature studies in Panama.* New York, London: D. Appleton and Company.
Cherrie, G. K. 1916. A contribution to the ornithology of the Orinoco region. *Bull. Brookl. Inst. Arts & Sci.* 2: 133a–374.
Clark, H. L. 1913. Notes on the Panamá Thrush-Warbler. *Auk* 30: 11–15.
du Noüy, L. 1947. *Human destiny.* London: Longmans, Green and Co.
Emlen, J. T. 1973. Territorial aggression in wintering warblers at Bahama agave blossoms. *Wilson Bull.* 85: 71–74.
ffrench, R. 1973. *A guide to the birds of Trinidad and Tobago.* Wynnewood, Pa.: Livingston Publishing Co.
Giles, L. 1948. *A gallery of Chinese immortals.* London: John Murray.
Goodwin, D. 1978. *Birds of man's world.* London: British Museum (Natural History).
Hartshorne, C. 1973. *Born to sing: An interpretation and world survey of bird song.* Bloomington: Indiana University Press.
Haug, M., and West, E. W. 1872. *The book of Arda Viraf.* Bombay and London.
Howell, T. R. 1964. Mating behavior of the Montezuma Oropendola. *Condor* 66: 511.
Howell, T. R., and Dawson, W. R. 1954. Nest temperatures and attentiveness in the Anna Hummingbird. *Condor* 56: 93–97.
Hudson, W. H. 1920. *Birds of La Plata.* 2 vols. London: J. M. Dent and Sons.
Huxley, J. 1923. *Essays of a biologist.* New York: Alfred A. Knopf.
Immelmann, K. 1966. Beobachtungen an Schwalbenstaren. *Jour. für Ornith.* 107: 37–69.
Kamil, A. C., and van Riper III, C. 1982. Within-territory division of foraging space by male and female Amakihi (*Loxops virens*). *Condor* 84: 117–119.
Kiltie, R. A., and Fitzpatrick, J. W. 1984. Reproduction and social organization of the Black-capped Donacobius (*Donacobius atricapillus*) in southeastern Perú. *Auk* 101: 804–811.

Lack, D. 1956. *Swifts in a tower*. London: Methuen & Co.
——. 1968. *Ecological adaptations for breeding in birds*. London: Methuen & Co.
Meyer de Schauensee, R. 1970. *A guide to the birds of South America*. Wynnewood, Pa.: Livingston Publishing Co.
Moehlman, P. D. 1980. Jackals of the Serengeti. *Natl. Geographic* 158: 840–850.
Moore, R. T. 1947. Habits of male hummingbirds near their nests. *Wilson Bull.* 59: 21–25.
Morton, E. S. 1976. Vocal mimicry in the Thick-billed Euphonia. *Wilson Bull.* 88: 485–487.
Noakes, D. L. G., and Barlow, G. W. 1973. Cross-fostering and parent-offspring responses in *Cichlasoma citrinellum* (Pisces, Cichlidae). *Z. Tierpsychol.* 33: 147–152.
Otto, M. C. 1949. *Science and the moral life*. New York: New American Library.
Regan, T. 1983. *The case for animal rights*. Berkeley and Los Angeles: University of California Press.
Remsen, J. V., Jr. 1976. Observations of vocal mimicry in the Thick-billed Euphonia. *Wilson Bull.* 88: 487–488.
Robins, C. R., and Heed, W. B. 1951. Bird notes from La Joya de Salas, Tamaulipas. *Wilson Bull.* 63: 263–270.
Russell, S. M. 1964. *A distributional survey of the birds of British Honduras*. American Ornithologists' Union: Ornith. Monograph 1.
Salvin, O., and Godman, F. D. 1879. *Biologia Centrali-Americana*. Vol. 1. London: Taylor & Francis.
Schäfer, E. 1954. Sobre la biología de *Colibri coruscans*. *Bol. Soc. Venezolana Cien. Nat.* 15: 153–162.
Skutch, A. F. 1931. The life history of Rieffer's Hummingbird (*Amazilia tzacatl tzacatl*) in Panama and Honduras. *Auk* 48: 481–500.
——. 1953. How the male bird discovers the nestlings. *Ibis* 95: 1–37, 505–542.
——. 1954. *Life histories of Central American birds*. Pacific Coast Avifauna 31. Berkeley: Cooper Ornithological Society.
——. 1965. Life history of the Long-tailed Silky-Flycatcher, with notes on related species. *Auk* 82: 375–426.
——. 1969a. *Life histories of Central American birds. III*. Pacific Coast Avifauna 35. Berkeley: Cooper Ornithological Society.
——. 1969b. A study of the Rufous-fronted Thornbird and associated birds. *Wilson Bull.* 81: 5–43, 123–139.
——. 1971. *A naturalist in Costa Rica*. Gainesville: University of Florida Press.
——. 1972. *Studies of tropical American birds*. Publ. Nuttall Ornith. Club 10. Cambridge, Mass.: Nuttall Ornithological Club.
——. 1976. *Parent birds and their young*. Austin: University of Texas Press.
——. 1977. *A bird watcher's adventures in tropical America*. Austin: University of Texas Press.
——. 1979. *The imperative call: A naturalist's quest in temperate and tropical America*. Gainesville: University Presses of Florida.
——. 1981. *New studies of tropical American birds*. Publ. Nuttall Ornith. Club 19. Cambridge, Mass.: Nuttall Ornithological Club.
——. 1983. *Nature through tropical windows*. Berkeley and Los Angeles: University of California Press.
Snow, B. K. 1974. Vocal mimicry in the Violaceous Euphonia, *Euphonia violacea*. *Wilson Bull.* 86: 179–180.
Snow, B. K., and Snow, D. W. 1971. The feeding ecology of tanagers and honeycreepers in Trinidad. *Auk* 88: 291–322.
Sturgis, B. B. 1928. *Field book of birds of the Panama Canal Zone*. New York and London: G. P. Putnam's Sons.
Tramer, E. J., and Kemp, T. R. 1980. Foraging ecology of migrant and resident warblers and vireos in the highlands of Costa Rica. In *Migrant birds in the Neotropics: Ecology, behavior, distribution, and conservation*, ed. A. Keast and E. S. Morton, pp. 285–296. Washington, D.C.: Smithsonian Institution Press.
Vaurie, C. 1980. Taxonomy and geographical distribution of the Funariidae (Aves, Passeriformes). *Bull. Amer. Mus. Nat. Hist.* 166: 1–357.
Vleck, C. M. 1981. Hummingbird incubation: Female attentiveness and egg temperature. *Oecologia* (Berlin) 51: 199–205.
Wagner, H. O. 1959. Beitrag zum Verhalten des Weissohrkolibris (*Hylocharis leucotis* Vieill.). *Zool. Jahrb. Syst.* 86: 253–302.
Wheeler, W. M. 1920. The Termitodoxa, or biology and society. *Scientific Monthly* (February).
Wilson, E. O. 1975. *Sociobiology: The new synthesis*. Cambridge: Harvard University Press, Belknap Press.

Index

Illustrations are indicated by **boldfaced** numbers.

Agouti, *Dasyprocta punctata,* 148, **86, 223**
Ahimsa (harmlessness): in Buddhism, 88; in China, 88; in Greco-Roman culture, 89–90; in Jainism, 87–88; as a perennial ideal, 91
Alchornea costaricensis (Euphorbiaceae), 84
Alder, *Alnus acuminata* (Betulaceae), 129
Amakihi, Common, *Hemignathus (Loxops) virens,* 110
Ani, Groove-billed, *Crotophaga sulcirostris,* 192–193, 197
  Smooth-billed, *C. ani,* 98
Animals: admirable traits of, 121–124; distorted by domestication, 124–125; exhibit some of man's finest attributes, 125
Animals and plants, treatment of: its pertinence to ethics, 85–87; in various cultures, 87–91
Annatto tree, *Bixa orellana* (Bixaceae), 116
Anticipatory food-bringing, by birds, 216
Anting, by birds, 96
Antshrike, Black-hooded, *Thamnophilus bridgesi,* 216, 217
  Slaty, *T. punctatus,* 218
Ant-Tanager, Red-crowned, *Habia rubica,* 216
Appreciation: binds things together ideally, 70–71; transcends biological insulation, 73; unification of mankind by shared, 71–73
Aracari, Collared, *Pteroglossus torquatus,* 193

Aragres, *Trigona* sp., 66
Arda Viraf, 88
Aristolochia grandiflora (Aristolochiaceae), 58
  pilosa, **60, 61**; pods and seeds of, 61–62; pollination of, 58–61
Aristotle, 21, 102, 221, 223
Asceticism: compared with conviviality, 156–159; and the conservation of nature, 157; and spirituality, 158–159
Attila, Bright-rumped, 207, 215

Baker, Herbert G., 64
Banana, *Musa sapientum* (Musaceae), 152–153
Bananaquit, *Coereba flaveola,* 40, 41, 210; usurps euphonia's dormitory, 204; visits *Mucuna* flowers, 64
Barbet, Prong-billed, *Semnornis frantzii,* fights with silky-flycatchers, 137–138
Barbthroat, Band-tailed, *Threnetes ruckeri,* 118
Barro Colorado Island, 56
Barva, Volcán, 129
Bat(s), reproduction of, 146, 147
  Sac-wing, *Saccopteryx bilineata,* 146
Beauty: of frugivorous and nectarivorous animals, 16–17; origins of, 14–22; promoted by constructive interactions between animals and plants, 15–16, 22
Beaver, family life of, 45
Beehler, Bruce M., 109–110
Bees, collect resin from clusia flowers, 66

Beetle, rove, or staphylinid, in *Aristolochia* flowers, 59–61, **60**
Beggar's Ticks, *Desmodium* spp. (Leguminosae), 64
Bent, Arthur Cleveland, 166
Biology, and ethics, 101–105
Birds: advantages of mating system of, 45–46; aggressiveness of nectarivorous, 109–110; conditions that favor monogamy in, 44; conscience in, 144; cooperative breeding in, 44–45, 166; effects of volcanic ash on, 131–132; eggs of, 146–147; emotions of parent, 211–221; equality of sexes in, 45; exemplify platonic love, 46; as indicators of nature's trend, 170–171; inefficient reproductive system of, 146; influences of weather on nestling periods of, 116–117; intelligence of, 221; mating systems of, 43–44, 45–46; mildness of frugivorous, 110; nest repair by, 27, 32; nests of, 142–143; parasitic, 46; protomorality of, 145–146; ways of feeding young by, 40–43. *See also* Dove; Castlebuilder; etc.
Bird watching, motives and rewards of, 167–171
Blackbird, Melodious, *Dives dives*, 100, 176, 215
Bladderworts, *Utricularia* spp. (Lentibulariaceae), 155
Blake, William, quoted, 90
Blechnum volubilis (Polypodiaceae), 150–151, **150**
Bluebirds, *Sialia* spp., 44, 193
Bobolink, *Dolichonyx oryzivorus*, 100, 202
Bowstring Hemp, *Sansevieria guineensis* (Liliaceae), 154
Bracken Fern, *Pteridium aquilinum* (Polypodiaceae), 150
Brush-Finch, Chestnut-capped, *Atlapetes brunneinucha*, 175, 176
  Yellow-throated, *A. gutturalis*, 172
Bryophyllum calycinum (*pinnatum*) (Crassulaceae), 154
Buddha, 103
Buddhism, 88
Bushtit, *Psaltriparus minimus*, nest building by, 69–70

Calabash tree, *Crescentia cujete* (Bignoniaceae), 113
Calliandra similis (Leguminosae), 191, 201
Canidae, cooperative breeding in, 45
Caracara, Yellow-headed, *Milvago chimachima*, 99
Carboniferous period, 15
Cardinal, Northern, *Cardinalis cardinalis*, 193
Caring: expansion of, 196–198; and education, 196; by human children, 194–196; by young birds and mammals, 192–194
Castlebuilder, Pale-breasted, *Synallaxis albescens*: appearance of, 36; distribution and habits of, 36; nesting of, 37–38; voice of, 36–38
  Rufous-breasted, *S. erythrothorax*: appearance of, 23; distribution and habitat of, 23, 38; nesting of, 23–29; nest repair by, 27; voice of, 23, 27–28
  Slaty, *S. brachyura*, **30**; appearance of, 29; distribution of, 29, 38; habits of, 29–30; voice of, 31
Castor-oil Plant, *Ricinus communis* (Euphorbiaceae), 152
Catbird, Gray, *Dumetella carolinensis*, 162, 176, 179, 218
Cecropia tree, *Cecropia* spp. (Moraceae), 84, 152, 201, 204
Cenozoic era, 15
Chakravarthy, A., 87
Chapman, Frank M., 77
Chat-Tanager, *Calyptophilus frugivorus*, 191
Cherrie, George K., 37
Chimpanzee, *Pan satyrus*, 195
China, attitude toward living things in ancient, 88
Chlorophonia(s), 199, 201, 204, 207
  Golden-browed, *Chlorophonia callophrys*, 140, 205
Citharexylum mocinnii (Verbenaceae), 129
Clark, H. L., 190
Clematis spp. (Ranunculaceae), 153
Clusia spp. (Guttiferae), **65**; pollination of, 64–67
  uvitana, 65–66
Coffee, synchronous flowering of, 67
Coleridge, Samuel Taylor, 90
Color vision, evolution of, 16–18
Colura spp. (Lejeuneaceae), 154
Conscience, adumbrations of, in birds, 144
Conservation: and asceticism, 156–159; discrimination in, 55–56; and ethics, 85
Cooperation, between diverse organisms, 55
Cooperative breeding, by birds and mammals, 44–45, 166
Cordillera de Talamanca, 188
Cornel, *Cornus disciflora* (Cornaceae), 129
Cosmos, as an ordered system, 181
Cotinga, Turquoise, *Cotinga ridgwayi*, 130
Cowbird, Giant, *Scaphidura* (*Psomocolax*) *oryzivora*: at oropendolas' nests, 79–80; fledgling of, 83
Cypress, Mexican, *Cupressus lusitanica* (Coniferae), 129

Darwin, Charles, 16, 18, 119
Dicranopteris pectinata (Gleicheniaceae), **7, 149**; scandent leaves of, 148–150
Distraction displays: and parental emotion, 218; by White-tipped Doves, 10–11, 12
Donacobius, Black-capped, *Donacobius atricapillus*, **161**; appearance of, 160; classification of, 166; cooperative breeding by, 166; food of, 160–162; nesting of, 163–166; voice and displays of, 160, 162–163
Dove, Rufous-naped, *Leptotila cassinii rufinucha*, 6
  White-fronted. *See* Dove, White-tipped

White-tipped, *L. verreauxi*, 7; defense of nest by, 11; distraction displays of, 10–11, 12, 218; distribution of, 6–8; nesting of, 8–13; resists snake, 11; roosting of, 12; voice of, 8
Dryopteris sp. (Polypodiaceae), 154
Du Noüy, Lecomte, 119
Dusky Titi, *Callicebus moloch*, 44
Duty, sense of, in birds, 144

Earth, fitness of, for a spiritual being, 3
Eggs, vulnerability of, 146
Egypt, attitude toward animals in ancient, 89
Einstein, Albert, 183
Elaenia, Lesser, *Elaenia chiriquensis*, 215, 216
  Mountain, *E. frantzii*, 137
  Yellow-bellied, *E. flavogaster*, 117
Emlen, John T., 110
Entada gigas (Leguminosae), 153
Ethics: and biology, 101–105; and conservation, 85
Euphonia, 199
Euphonia(s): appearance of, 199–201; conflicts of, 210; food of, 201–202; nesting of, 204–210; nuptial feeding by, 202; sleeping by, 203–204; vocal mimicry of, 203; voice of, 202–203, 210
  Blue-hooded, *Euphonia elegantissima*, 201, 202, 204
  Golden-sided, *E. cayennensis*, 201
  Jamaican, *Pyrrhuphonia (Euphonia) jamaica*, 204
  Olive-backed, *Euphonia gouldi*, 201, 202, 204, 205
  Purple-throated, *E. chlorotica*, 205
  Spotted-crowned, 200. *See also* Euphonia, Tawny-bellied
  Tawny-bellied, *E. imitans*, 201–210 *passim*, **200**
  Thick-billed, *E. laniirostris*, 203, 205
  Trinidad, *E. trinitatis*, 202
  Violaceous, *E. violacea*, 203, 205
  White-vented, *E. minuta*, 202–210 *passim*, 221
  Yellow-crowned, *E. luteicapilla*, 204, 205, 207, 209
  Yellow-throated, *E. hirundinacea*, 199, 205, 206, 207, 209
Eurya theoides (Theaceae), 129
Evolution: chance in, 182–184; efficiency of, 147
Eye, evolution of, 3–4

Fairy, Purple-crowned, *Heliothryx barroti*, 109, **48, 49**; appearance of, 47; flycatching by, 50; a nectar-thief, 49–50; nesting of, 50–51; visits *Mucuna* flowers, 64
Faith, 143–144
Fer-de-lance, *Bothrops atrox*, 52, 56
Fern(s): climbing, 148–151, **150**; walking, 153–154, **149**
  Bladder, *Cystopteris* spp. (Polypodiaceae), 153
  Star, *Hemionitis* sp. (Polypodiaceae), 154
  Walking Leaf, *Camptosorus rhizophyllus* (Polypodiaceae), 153
Fernández Soto, Rafael Angel, 129
ffrench, Richard, 37, 205
Firewood-gatherer, *Anumbius annumbi*, 24
Flame-of-the-Forest tree, *Spathodea campulata* (Bignoniaceae), 84, 152
Flies, pollinate flowers of *Aristolochia*, 59–61
Flower-piercers, *Diglossa* spp., 49
Flycatcher(s) (Tyrannidae), 207, 214, 215
  Boat-billed, *Megarhynchus pitangus*, 214, 215, 216–217
  Dusky-capped, *Myiarchus tuberculifer*, 215
  Golden-bellied, *Myiodynastes hemichrysus*, 214
  Gray-capped, *Myiozetetes granadensis*, 33, 214, 215, 217, 219–221
  Great Crested, *Myiarchus crinitus*, 26
  Great Kiskadee. *See* Kiskadee, Great
  Piratic, *Legatus leucophaius*, 215
  Royal, *Onychorhynchus coronatus*, 205
  Sulphur-bellied, *Myiodynastes luteiventris*, 214
  (Tyrannulet), Torrent, *Serpophaga cinerea*, 52, **53**
  Vermilion-crowned, *Myiozetetes similis*, 33, 163, 214, 215, 217, 218–219
Forest, tropical rain: advancement of knowledge in, 56; as a maximum expression of the creative energy, 53–54
Frullania spp. (Jungermanniaceae), 153, 203
Fuchsia, Tree, *Fuchsia arborescens* (Onagraceae), 129

Gallinule(s), 193
  Common, or Moorhen, *Gallinula chloropus*, cooperative breeding by, 44
  Purple, *Porphyrula martinica*, 163; cooperative breeding by, 44
Gibbon, **19**
Giles, Lionel, 88
God: as Divine Astrophysicist, 181–182; and religion, 184–185; as Supreme Chemist, 182
Goodness: as an ideal, 104–105; and knowledge, 104; meaning of, 101
Goodwin, Derek, 110
Grackle, Great-tailed, *Quiscalus (Cassidix) mexicanus*, 96
Grosbeak, Black-faced, *Caryothraustes poliogaster*, 207
  Black-headed, *Pheucticus melanocephalus*, 215
  Blue-black, *Cyanocompsa cyanoides*, 215, 216
  Rose-breasted, *Pheucticus ludovicianus*, 215
Ground-Dove, Blue, *Claravis pretiosa*, 8
Growth, as a mode of harmonization, 57
Guarea rhopalocarpa (Meliaceae), indeterminate growth of leaves of, 151–152

Hairy Birthwort, *Aristolochia pilosa* (Aristolochiaceae), **60, 61**; pods and seeds of, 61–62; pollination of, by trapped flies, 58–61

Harmlessness. See Ahimsa
Harmonization: revealed in growth, 57; significance of, 186–187; a universal process, 185
Hartshorne, Charles, 213
Haug, Martin, and E. W. West, 88
Hawk, Red-tailed, *Buteo jamaicensis*, 140
Hermit, Little, *Phaethornis longuemareus*, 49, 210; visits *Mucuna* flowers, 64
Honeycreeper, Green, *Chlorophanes spiza*, 110
Honeyeater, Singing, *Meliphaga virescens*, 110
House-Wren, Southern, *Troglodytes musculus*, 38, 193–194, 219
Howell, Thomas R., 79; and W. R. Dawson, 114
Hudson, W. H., 24
Humboldt, Alexander von, 58
Hummingbird, Anna's, *Calypte anna*, 114
　Charming, *Amazilia decora*, 47
　Little Hermit. See Hermit, Little
　Long-billed Starthroat. See Starthroat, Long-billed
　Purple-crowned Fairy. See Fairy, Purple-crowned
　Rufous-tailed, *Amazilia tzacatl*, 47, 108–109, 110, 117, 118
　Scaly-breasted, *Phaeochroa cuvierii*, 49, **107**; bathing by, 110; departure from nest of, 116–117; duration of parental care by, 117–118; foraging by, 109; nesting of, 110–115; nesting success of, 118; singing assemblies of, 106–109
　Snowy-bellied, *Amazilia edward*, 47
　White-eared, *Hylocharis leucotis*, 49, 113, 114, 118
　White-necked Jacobin. See Jacobin, White-necked
　Wine-throated, *Atthis ellioti*, 108
Huxley, Julian, 213

Ideals, as receding goals, 91, 146
Immelmann, Klaus, 145–146
Injury-feigning. See Distraction displays
Insulation, as a basic condition of life, 68–69
Irazú, Volcán: birds on, 127; effects of eruption on birds, 131–132; eruption of, 131
Isaiah, quoted, 89
Israel, attitude toward animals in ancient, 89

Jacamar, Pale-headed, *Brachygalba goeringi*, 219, **220**
　Rufous-tailed, *Galbula ruficauda*, 219
Jacaranda, Large-leaved, *Jacaranda copaia* (Bignoniaceae), 152
Jackal, Golden, *Canis aureus*, 45
　Silver-backed, *C. mesomelas*, 45
Jacobin, White-necked, *Florisuga mellivora*, **212**
Jaguar, *Felis onca*, 52
Jainism, 87–88
Jay, Brown, *Cyanocorax* (*Psilorhinus*) *morio*, 82, 134, 140–141, 193, 197
　Florida Scrub, *Aphelocoma coerulescens*, cooperative breeding by, 44
　Gray-breasted (Mexican), *A. ultramarina*, cooperative breeding by, 44
Jeans, Sir James, 181
Jesus, 103

Kalidasa, 90
Kamil, Alan C., and Charles van Riper III, 110
Kestrel, American, *Falco sparverius*, 97, 98
Kiltie, Richard A., and John W. Fitzpatrick, 166
Kingbird, Tropical, *Tyrannus melancholicus*, 130
Kiskadee, Great, *Pitangus sulphuratus*, 80, 205, 217
Kite, Double-toothed, *Harpagus bidentatus*, 42

Lack, David, 43, 117
La Giralda, Finca, 129
Lao-tzu, 69, 103
La Selva, Finca, 56
Leaves: climbing, 148–151; as food traps, 154–155; indeterminate growth of, 148–152; reproduction by, 153–154; as substitutes for branches, 152; as substitutes for trunks, 152; as tendrils, 153; water storage by, 153
Leptochilus cladorrhizans (Polypodiaceae), 154
Life: evolution of, 182–184; excessive abundance of, 185; as goal of matter, 185
Light: properties of, 1–5; as a unifying agent, 4
Lily, Climbing, *Gloriosa superba* (Liliaceae), 153
Liverworts, 153, 154
Lizards, 34–35
Lucia (lizard), *Leiolopisma cherriei*, 35

Madera Negra, *Gliricidia sepium* (Leguminosae), 47, 50, 219
Mahavira, 103
Mammals: cooperative breeding of, 45; mating systems of, 43–44; method of feeding young of, 43
Man: aesthetic adaptation of, 20–21; breeding system of, 45–46; contribution to cosmic perfection of, 20; development of aesthetic sensibility of, 17–22; evolution of, 119; loneliness of, 186; rapprochement to nature by, 52–57; repudiation of animal ancestry by, 119–121; sources of strength of, 186–187; valuable traits shared with other animals by, 121–124
Marica caerulea (Iridaceae), synchronous flowering of, 67
Martin, House, *Delichon urbica*, 44
Maxillaria sp. (Orchidaceae), pollination of, 59
Maximum elaboration and elevation: examples of, 91–93; principle of, 91
Megapodes, incubation mounds of, 46
Mencius, 88
Meyer de Schauensee, Rodolphe, 94
Mica (snake), *Spilotes pullatus*, 80

Miconia spp. (Melastomaceae), synchronous flowering of, 67
Mistletoes (Loranthaceae), as food of euphonias, 201, 209
Mockingbird, Blue-and-White, *Melanotis hypoleucus*, 191, **189**; appearance of, 174; food of, 174; nesting of, 175–180; plumage changes of, 180; problem-solving by, 177–179; voice of, 172–180 *passim*
Mockingthrush, Black-capped. *See* Donacobius, Black-capped
Moehlman, Patricia D., 45
Monogamy: conditions that favor, 44–45, 147; in man, 45; prevalence of, in birds and mammals, 43–44
Moore, Robert T., 113
Morality, innate foundation of, 145–146
Morton, Eugene S., 203
Motmot, Blue-diademed (Blue-crowned), *Momotus momota*, 118
Mucuna holtoni (Leguminosae), **63**; pollination of, 62–64

Natural selection, 183–184
Nature: conflicts in the study of, 101–105; intangible values from, 57; man's bonds with, 101, 105; pacification of, 55; penetrating the surface of, 56–57; rapprochement of man to, 52–57
Nature reserves, need of, 56
Neanderthal man, 186
Nemi Natha, 87
Nightingale-Thrush, Orange-billed, *Catharus aurantiirostris*, 29–31, 217
  Ruddy-capped, *C. frantzii*, 175, 176
Noakes, David L. G., and George W. Barlow, 197

Ojo de Buey, *Mucuna* spp. (Leguminosae), 62
Oriole, Baltimore (Northern), *Icterus galbula*, 47
  Black-cowled, *I. dominicensis*, 217
  Orange-crowned, *I. auricapillus*, 100
  Orchard, *I. spurius*, 47
  Yellow, *I. nigrogularis*, 100
  Yellow-tailed, *I. mesomelas*, 174, 176, 215
Oriole-Blackbird, *Gymnomystax mexicanus*, **95**; anting by, 96; food of, 99; habits and nesting of, 94–100
Oropendola, Chestnut-headed, *Psarocolius (Zarhynchus) wagleri*, 77, 84
  Crested, *P. decumanus*, 100
  Montezuma, *P. (Gymnostinops) montezuma*, **75**; distribution of, 83; food of, 84; nesting colony of, 74–83; relations of, with Giant Cowbirds, 79–80, 83; roosting of, 84
Otto, Max, quoted, 121
Ovenbird family (Furnariidae), 38–39

Paraulata de Agua, 162
Pauraque, Common, *Nyctidromus albicollis*, 81

Peccary, White-lipped, *Tayassu pecari*, 52
Pejibaye palm, *Bactris utilis*, 12, 83
Pelican, Pink-backed, *Pelecanus rufescens*, 42
Peppershrike, Rufous-browed, *Cyclarhis gujanensis*, 165, 215
Phainopepla, *Phainopepla nitens*, 127
Phelps, Kathleen Deery de, 94
Piculet, Olivaceous, *Picumnus olivaceus*, method of feeding young, 41
Pigeon, Short-billed, *Columba nigrirostris*, 78
Piper spp. (Piperaceae), 201
Plutarch, 90
Poró, Orange, *Erythrina poeppigiana* (Leguminosae), 84
  Red, *E. berteroana*, 49, 64, 109, **48**
Porphyry, 90
Predation, 182, 184
Primates: colors of, 18; evolution of, 17–18; monogamy in, 44
Protium sp. (Burseraceae), 201
Protomorality, of birds, 145–146
Ptilogonatidae, 127
Puma, *Felis concolor*, 52
Pythagoras, 89

Quail-Dove, Ruddy, *Geotrygon montana*, 10
Queo, *Rhodinocichla rosea*, **173**; appearance of, 188–189; classification of, 189, 191; distribution of, 189; food of, 190; nest of, 191; voice of, 188, 190, 191
Quetzal, Resplendent, *Pharomachrus mocinno*, 127

Redstart, Slate-throated, *Myioborus miniatus*, 177
Regan, Tom, 103
Religion, 184–185
Remsen, J. V., Jr., 203
Rhipidopteris peltata (Polypodiaceae), 206
Rhodinocichla rosea, 189–191
Río Buena Vista, 188
Río Peñas Blancas, 52, 153
Robin. *See* Thrush
Robins, C. Richard, and William B. Heed, 204
Rose-apple tree, *Eugenia jambos* (Myrtaceae), 216
Rotifers, in leaves of *Frullania*, 153
Russell, Bertrand, 186
Russell, Stephen M., 205

St. Francis of Assisi, 90
St. Paul, 103
Salpichlaena (Blechnum) volubilis (Polypodiaceae), **150**; scandent fronds of, 150–151
Saltator, Buff-throated, *Saltator maximus*, 118, 165, 202, 214, **120**
  Streaked, *S. albicollis*, 165, 201
Salvin, Osbert, 175
Sansevieria guineensis (Liliaceae), propagation of, 154
Schäfer, Ernst, 113
Schwartz, Paul, 99, 191

Seedeater, Variable, *Sporophila aurita,* 190, 217–218
    White-collared, *S. torqueola,* 217
Seneca, Lucius Annaeus, quoted, 90
Sexual reproduction, compared to gambling, 183
Sexual selection, 16–17
Shelley, Percy Bysshe, quoted, 90
Silky-Flycatcher, Black-and-Yellow, *Phainoptila melanoxantha,* 127
    Gray, *Ptilogonys cinereus,* 127
    Long-tailed, *P. caudatus,* **128**; appearance of, 127; enemies of, 140–141; fight of, with Prong-billed Barbet, 137–138; food of, 129–130; habitat of, 129; nesting of, 132–140; nesting success of, 140; social habits of, 130–131; voice of, 130
Snake: attacks dove's nest, 11; at oropendolas' tree, 80
Snow, Barbara K., 203; and David W. Snow, 202
Souroubea guianensis (Marcgraviaceae), 201
Space, mystery of, 2–3
Sparrow, Black-striped, *Arremonops conirostris,* 148, 214
Sparrow Hawk. *See* Kestrel, American
Spencer, Herbert, 145
Sphagnum spp. (Sphagnaceae), 153
Spirituality, 158–159
Starthroat, Long-billed, *Heliomaster longirostris,* 49, 109, 118, **49**
Stoics, attitude toward animals of, 90
Sturgis, Bertha Bemis, 51
Sundews, *Drosera* spp. (Droseraceae), 154
Swallow(s), 193
    Barn, *Hirundo rustica,* juvenile helpers of, 44
Sweet Pea, *Lathyrum odoratum* (Leguminosae), 153
Swift, European (Common), *Apus apus,* 117
Synallaxis, 38

Tanager(s), 199, 202; mildness of, at fruiting trees, 110
    Blue (Blue-gray), *Thraupis episcopus,* 201–202
    Golden-masked (Golden-hooded), *Tangara larvata,* 193, 194, 195, 202
    Scarlet-rumped, *Ramphocelus passerinii,* 148, 207, 209, 214, 218, 221
Taoism, 88
Tennyson, Alfred, quoted, 90–91
Tern, Sooty, *Sterna fuscata,* 42
Thornbird, Rufous-fronted, *Phacellodomus rufifrons,* 39, 100, 205
Thrasher, Brown, *Toxostoma rufum,* 179
    Crissal, *T. dorsale,* 166
    Palmer's (Curve-billed), *T. curvirostre palmeri,* 166
Thrush, Bare-eyed, *Turdus nudigenis,* 165
    Black, *T. infuscatus,* 174, 176
    Garden (Clay-colored), *T. grayi,* 84, 214, 217
    Mountain, *T. plebejus,* 137

Thrush-Tanager, Rosy, 189, **189**. *See also* Queo
Tibet, animals in, 88
Tillandsia usneoides (Bromeliaceae), 204
Tinamou, Great, *Tinamus major,* 78
Tityra, Black-crowned, *Tityra inquisitor,* 219
Tody-Flycatcher, Back-fronted (Common), *Todirostrum cinereum,* 108, 207
Toucanet, Emerald, *Aulacorhynchus prasinus,* 140
Tramer, Elliot J., and Thomas R. Kemp, 110
Trigona bees, as pollinators, 66, 67
Troupial, *Icterus icterus,* 100, 203
Tyranniscus, Paltry, *Zimmerius (Tyranniscus) vilissimus,* 210, 216

Universe: beauty of, 191; striving of, to enhance existence, 4–5, 54–55, 186–187
Utricularia endresii (Lentibulariaceae), 155

Values, universal striving to realize, 3–5
Vaurie, Charles, 38
Vegetarian diet, 92–93
Venus's Flytrap, *Dionaea muscipula* (Droseraceae), 154–155
Violet-ear, Sparkling, *Colibri coruscans,* 113
Vireo, Warbling, *Vireo gilvus,* 215
    Yellow-green, *V. flavoviridis,* 203
    Yellow-throated, *V. flavifrons,* 215
Virgin's Bower, *Clematis* spp. (Ranunculaceae), 153
Visual image, inversion of, 4
Vleck, Carol M., 51

Wagner, Helmuth O., 113
Wallace, Alfred Russel, 119
Warbler, Black-cheeked, *Basileuterus melanogenys,* 131
    Cape May, *Dendroica tigrina,* 110
    Flame-throated, *Parula (Vermivora) gutturalis,* 131, 137
    Mourning, *Oporornis philadelphia,* 36
    Palm, *Dendroica palmarum,* 110
    Tennessee, *Vermivora peregrina,* 110
    Wilson's, *Wilsonia pusilla,* 137
Waxwing, Cedar, *Bombycilla cedrorum,* 130, 131
Welfare state, 144–145
Welwitschia mirabilis (Gnetaceae), 152
Wheeler, William Morton, quoted, 146
Whip-poor-will, *Caprimulgus vociferus,* 174
Wilson, Edward O., 44
Winter's Bark, *Drimys winteri* (Winteraceae), 129
Woodpecker(s): cooperative breeding in, 44; methods of feeding young, 41
    Acorn, *Melanerpes formicivorus,* 44
    Black, *Dryocopus martius,* 41
    Green, *Picus viridis,* 41
    Pileated, *Dryocopus pileatus,* 41
    Red-cockaded, *Picoides borealis,* 44

Wood-Quail, Marbled, *Odontophorus gujanensis*, 219
Wood-swallows (Artamidae), 145
Wood-Wren, Highland (Gray-breasted), *Henicorhina leucophrys,* 215
Woolman, John, quoted, 91
World process, as movement to enhance existence, 4–5

Wren, Banded-backed, *Campylorhynchus zonatus,* 193
  Spotted-breasted, *Thryothorus maculipectus,* 114

Zarathustra, 89
Zoroastrians, attitude toward animals and plants of, 88